AF828852

Experiments and Numerical Simulations of Turbulent Combustion of Diluted Sprays

ERCOFTAC SERIES

VOLUME 19

Series Editor

Bernard Geurts
University of Twente Fac. Mathematical Sciences, Enschede, The Netherlands

Aims and Scope of the Series

ERCOFTAC (European Research Community on Flow, Turbulence and Combustion) was founded as an international association with scientific objectives in 1988. ERCOFTAC strongly promotes joint efforts of European research institutes and industries that are active in the field of flow, turbulence and combustion, in order to enhance the exchange of technical and scientific information on fundamental and applied research and design. Each year, ERCOFTAC organizes several meetings in the form of workshops, conferences and summerschools, where ERCOFTAC members and other researchers meet and exchange information.

The ERCOFTAC Series will publish the proceedings of ERCOFTAC meetings, which cover all aspects of fluid mechanics. The series will comprise proceedings of conferences and workshops, and of textbooks presenting the material taught at summerschools.

The series covers the entire domain of fluid mechanics, which includes physical modelling, computational fluid dynamics including grid generation and turbulence modelling, measuring-techniques, flow visualization as applied to industrial flows, aerodynamics, combustion, geophysical and environmental flows, hydraulics, multiphase flows, non-Newtonian flows, astrophysical flows, laminar, turbulent and transitional flows.

For further volumes:
http://www.springer.com/series/5934

Bart Merci • Eva Gutheil
Editors

Experiments and Numerical Simulations of Turbulent Combustion of Diluted Sprays

TCS 3: Third International Workshop on Turbulent Spray Combustion

Springer

Editors
Bart Merci
Flow, Heat and Combustion Mechanics
Ghent University
Ghent
Belgium

Eva Gutheil
Interdisciplinary Center for Scientific Computing
Heidelberg University
Heidelberg
Germany

ISSN 1382-4309
ISBN 978-3-319-04677-8 ISBN 978-3-319-04678-5 (eBook)
DOI 10.1007/978-3-319-04678-5
Springer Cham Heidelberg New York Dordrecht London

Library of Congress Control Number: 2014935894

© Springer International Publishing Switzerland 2014
This work is subject to copyright. All rights are reserved by the Publisher, whether the whole or part of the material is concerned, specifically the rights of translation, reprinting, reuse of illustrations, recitation, broadcasting, reproduction on microfilms or in any other physical way, and transmission or information storage and retrieval, electronic adaptation, computer software, or by similar or dissimilar methodology now known or hereafter developed. Exempted from this legal reservation are brief excerpts in connection with reviews or scholarly analysis or material supplied specifically for the purpose of being entered and executed on a computer system, for exclusive use by the purchaser of the work. Duplication of this publication or parts thereof is permitted only under the provisions of the Copyright Law of the Publisher's location, in its current version, and permission for use must always be obtained from Springer. Permissions for use may be obtained through RightsLink at the Copyright Clearance Center. Violations are liable to prosecution under the respective Copyright Law.
The use of general descriptive names, registered names, trademarks, service marks, etc. in this publication does not imply, even in the absence of a specific statement, that such names are exempt from the relevant protective laws and regulations and therefore free for general use.
While the advice and information in this book are believed to be true and accurate at the date of publication, neither the authors nor the editors nor the publisher can accept any legal responsibility for any errors or omissions that may be made. The publisher makes no warranty, express or implied, with respect to the material contained herein.

Printed on acid-free paper

Springer is part of Springer Science+Business Media (www.springer.com)

Preface

This book reflects the results of the 2nd and 3rd International Workshops on Turbulent Spray Combustion held in Chia Laguna (Sardinia) in 2011 and in Heidelberg (Germany) in 2012. The focus of these 'TCS' workshops is on reporting the progress of experimental and numerical techniques in two-phase flows, with emphasis on spray combustion. The motivation for studies in this area is that knowledge of the dominant phenomena and their interactions in such flow systems is essential for the development of predictive models and their use in combustor and gas turbine design. This necessitates the development of accurate experimental methods and numerical modelling techniques. The workshops aimed at providing an opportunity for experts and young researchers to present the state-of-the-art, discuss new developments or techniques and exchange ideas in the areas of experimentations, modelling and simulation of reactive multiphase flows.

The first chapter reflects on relevant issues of flame structure, auto-ignition and atomization with reference to well-characterized burners, to be implemented by modellers with relative ease. The second chapter presents an overview of first simulation results on target test cases, developed at the occasion of the 1st International Workshop on Turbulent Spray Combustion (Corsica 2009). In the third chapter, evaporation rate modelling aspects are covered, while the fourth chapter deals with evaporation effects in the context of flamelet models. In chapter five, LES simulation results are discussed for variable fuel and mass loading. Chapter six, on PDF modelling, makes the book complete.

In short, we believe that these contributions individually and together are highly valuable for the research community in this field. They allow experts and young scientists to gain more in-depth insight into some of the many aspects of dilute turbulent spray combustion, including state-of-the-art experiments and modelling efforts.

It is our intention to continue our efforts in bringing together experimentalists and modellers of spray combustion. In subsequent workshops, we intend to continue efforts on the 'target test cases', for which experimental databases have by now been made available and on which modelling and numerical algorithm issues are being tested. The definition and elaboration of other test cases is also an option for future studies.

When you are reading this book, TCS 4 will already have taken place in Cesme (Turkey) in 2013. We hope that this book may encourage you to participate in this quest in the future. You will find all information on http://www.tcs-workshop.org/

Sincerely,

Bart Merci and Eva Gutheil, Editors, On behalf of the TCS organizing committee.

Contents

Contributor

Mouldi Chrigui Department of Mechanical Engineering, Institute for Energy and Powerplant Technology, Technische Universität Darmstadt, Petersenstr. 30, Darmstadt, Germany

Research Unit Materials, Energy and Renewable Energies, University of Gafsa, Gafsa, Tunisia

Fernando Sacomano Department of Mechanical Engineering, Institute for Energy and Powerplant Technology, Technische Universität Darmstadt, Petersenstr. 30, Darmstadt, Germany

James D. Gounder Institut für Verbrennungstechnik, Deutsches Zentrum für Luft- und Raumfahrt, Germany

Eva Gutheil Interdisciplinary Center for Scientific Computing, University of Heidelberg, Heidelberg, Germany

Colin R. Heye Department of Aerospace Engineering and Engineering Mechanics, The University of Texas at Austin, Austin, Texas, USA

Yong Hu Interdisciplinary Center for Scientific Computing, University of Heidelberg, Heidelberg, Germany

Rana M. Humza Interdisciplinary Center for Scientific Computing, University of Heidelberg, Heidelberg, Germany

Agisilaos Kourmatzis School of Aerospace, Mechanical and Mechatronic Engineering, University of Sydney, NSW, Australia

School of Aerospace, Mechanical and Mechatronic Engineering, The University of Sydney, Darlington, Australia

Assaad R. Masri School of Aerospace, Mechanical and Mechatronic Engineering, University of Sydney, NSW, Australia

School of Aerospace, Mechanical and Mechatronic Engineering, The University of Sydney, Darlington, Australia

Hernan Olguin Interdisciplinary Center for Scientific Computing, University of Heidelberg, Heidelberg, Germany

William O'Loughlin Institut für Verbrennungstechnik, Deutsches Zentrum für Luft- und Raumfahrt, Germany

Venkat Raman Department of Aerospace Engineering and Engineering Mechanics, The University of Texas at Austin, Austin, Texas, USA

Amsini Sadiki Department of Mechanical Engineering, Institute for Energy and Powerplant Technology, Technische Universität Darmstadt, Petersenstr. 30, Darmstadt, Germany

From Dilute to Dense Turbulent Sprays: Combustion, Auto-Ignition and Atomization

Assaad R. Masri, Agisilaos Kourmatzis, William O'Loughlin and James D. Gounder

Abstract In line with the general theme of the International Workshop series on Turbulent Combustion of Sprays (TCS), this Chapter addresses relevant issues of turbulent flame structure, auto-ignition and atomization with reference to well-characterized burners that could be implemented by modelers with relative ease. The discussion of turbulent combustion is limited to dilute sprays stabilized on a simple piloted burner. Attention is shifted to the structure of the reaction zones and the challenges of computing chemical composition of flames of different fuels. Another section is dedicated to studying auto-ignition of turbulent dilute spray flames as observed in a hot vitiated co-flow. A common feature to all liquid fuels studied here is the presence of ignition kernels which grow (and sometimes extinguish) to induce flaming combustion further downstream. It is noted that this downstream region is responsible for the bulk of heat release and its local compositional structure depends on the parent fuel.

The last section focuses on non-reacting dense sprays with a view to shed some light on the effects of turbulence on secondary atomization. Two configurations are discussed: one for droplets flowing in a pipe and the other uses dense sprays generated with an air assisted atomizer. The carrier velocity is changed in both configurations to induce turbulence which affects atomization as clearly demonstrated by the reported measurements. While this chapter is not necessarily a roadmap for the efforts of future TCS workshops, the topics addressed here form a logical progression that is somewhat commensurate with the increasing order of complexity generally sought for computations.

A. R. Masri (✉) · A. Kourmatzis
School of Aerospace, Mechanical and Mechatronic Engineering,
University of Sydney, NSW 2006, Australia
e-mail: assaad.masri@sydney.edu.au

W. O'Loughlin · J. D. Gounder
Institut für Verbrennungstechnik, Deutsches Zentrum für Luft- und Raumfahrt,
Stuttgart 70569, Germany

B. Merci, E. Gutheil (eds.), *Experiments and Numerical Simulations of Turbulent Combustion of Diluted Sprays,* ERCOFTAC Series 19, DOI 10.1007/978-3-319-04678-5_1,
© Springer International Publishing Switzerland 2014

1 Introduction

Renewed interest in sprays is driven by a number of factors including the advent of novel, non-intrusive diagnostic tools and immense computational power that enable a fresh and more in-depth look at this complex topic. Another key driver is the necessity to develop clean alternatives to fossil-based fuels, such as biofuels and biodiesels. Biofuels are produced predominantly from biomass and, while they may be generated in gaseous or solid forms, the bulk of the production is liquid bio-alcohols and biodiesels [1–3]. Biodiesels form an important class of renewable biofuels that may be used directly in engines as blends or even as pure fuels. They may be alkyl esters of fatty acids extracted from a variety of feedstocks including vegetable oils, food waste, animal fats, or algae. Alternatively, biodiesels or bio-kerosene may be synthetic, gas-to-liquid (GTL) fuels produced from natural gas via Fischer-Tropsch processes [4]. GTL liquids consist of long-chain paraffin molecules and have negligible amounts of poly-aromatics and sulfur [5].

Biofuels bring additional complexities that compound the long term challenges associated with the utilization of sprays. The chemical kinetics of oxygenated fuels and their impact on engine performance, as well as on the formation of pollutants including particles, are subjects of current intense research [6–17]. These issues add to an already long list of outstanding sub-processes common to all liquid fuels and include (i) primary and secondary atomization, (ii) coalescence and dispersion, (iii) droplet evaporation and interaction with the surrounding turbulence and heat release, and (iv) droplet combustion [18], [19]. In parallel with many existing conferences and journals that report research advances on sprays, two international workshop series are bringing together numerical and experimental researchers to catalyze advances by focusing on well-documented spray flows representative of specific issues. The Turbulent Combustion of Spray Workshops (TCS-series) has, to date, addressed dilute sprays [20] while the Engine Network Group (ENG) has focused on high pressure, dense sprays that are more representative of engine conditions [21].

The utilization of sprays in engines is a very broad topic that, while related to the thesis of this chapter, remains well outside the scope of its limited coverage. The focus here is on the understanding of selected basic processes in dilute and dense sprays. More specifically, issues of atomization, auto-ignition and combustion of turbulent sprays are addressed with respect to specific burners and atomizers developed as well-designed platforms that can be adopted by modelers with relative ease. Atomization, both primary [22], [23] and secondary [24], [25], involves the processes of breaking up the liquid into ligaments and then smaller and smaller droplets which subsequently vaporize. Detailed measurements in the primary atomization regions of sprays are scarce and diagnostic techniques, such as ballistic imaging [26], [27] and X-Ray [28–30] methods are gradually evolving with promise to enable measurements of ligaments and the rate of mass loss from liquid cores. Knowledge of secondary atomization is more advanced such that modes of secondary break-up are well-correlated with relevant dimensionless parameters [24]. The effect of turbulence on both primary and secondary atomization remains, however, an area of importance that has not yet been thoroughly addressed.

Modes of auto-ignition in the presence of spays have recently been investigated using two configurations similar to those commonly used to study auto-ignition of gases. One configuration involves hot vitiated coflows [31] while the other uses heated air [32]. Simulations of ignition in sprays [33–35] reveal similarities with gaseous flames in that auto-ignition kernels prevail upstream of the main flaming region although the most ignitable mixture is affected by the presence of droplets. Measurements of turbulent, dilute spray jets auto-igniting in hot vitiated coflows are reported here to further elaborate on this topic which will form a challenge for modelers particularly in light of the strong dependence of auto-ignition on chemical kinetics.

The title of this chapter lists the three processes covered here in the reverse order of occurrence in flames but in the correct chronological order in which advances are made. The first three TCS workshops have focused on dilute spray jets and flames and issues of modeling droplet dynamics, evaporation processes as well as mean temperature fields. Details of such recent advances are covered in Chaps. 2, 5 and 6 of this book. This chapter attempts to probe additional aspects by addressing the compositional structure of spray flames as well as the nature of auto-ignition in dilute sprays. A new atomizer is also introduced and the issues of primary as well as secondary atomization are discussed with respect to the effects of turbulence and the physical properties of the sprays on these processes. It is envisaged that such topics would form the logical progression in focus for upcoming TCS workshops.

The chapter is organized as follows: Section 2 covers issues of the compositional structure of turbulent pilot-stabilized dilute spray flames. Section 3 describes the auto-ignition burner and discusses results with respect to auto-ignition modes and jet structure. Section 4 introduces two configurations, a simple pipe flow as well as an air-blast atomizer, for studying the effects of turbulent fluctuations on secondary atomization in non-reacting sprays. Section 5 closes with some concluding remarks.

2 Composition Fields of Dilute Spray Flames

The burner addressed in this section is the piloted spray burner which is already established as a model problem for the TCS workshops [20].

2.1 Burner and Flames

Full details of the piloted spray burner and the stabilized flames can be found elsewhere [36], [37] and only a brief description is given here. The spray is generated using a Sono-Tek ultrasonic nebulizer located 215 mm upstream and is advected to the burner's exit plane with a co-flowing stream of carrier air. The central fuel tube, with an inner diameter of 10.5 mm is surrounded by an annular pilot flame holder, which is 25 mm in diameter. A co-flow of diameter 104 mm surrounds the burner

Table 1 Initial conditions for the acetone and ethanol flames

Reacting Spray Cases	*1*	*2*	*3*	*4*	*5*	*7*
Bulk Jet Velocity U_{jet} (m/s)	24	36	24	24	48	60
Carrier mass flow rate (g/min)	150	225	150	150	301	376
Liquid fuel injection rate (g/min)	75	75	45	23.4	75	75
Overall equivalence ratio, $\Phi_{overall}$	4.7	3.2	2.9	1.5	2.4	1.9
Reacting Spray Cases—Acetone	*AcF 1*	*AcF 2*	*AcF 3*	*AcF 4*	*AcF 5*	*AcF 7*
Measured liquid flow at exit (g/min)	18.0	23.9	15.9	7.2	27.8	31.1
Vapor fuel flow at jet exit (g/min)	57.0	51.1	29.1	16.2	47.2	43.9
Equivalence ratio at jet exit, Φ_{exit}	3.6	2.2	1.8	1.0	1.5	1.1
Reacting Spray Cases—Ethanol	*EtF 1*	*EtF 2*	*EtF 3*	*EtF 4*	*EtF 5*	*EtF 7*
Measured liquid flow at exit (g/min)	45.7	66.6	30.7	14.5	70.1	73.0
Vapor fuel flow at jet exit (g/min)	29.3	8.4	14.3	8.9	4.9	2.0
Equivalence ratio at jet exit, Φ_{exit}	1.75	0.34	0.85	0.53	0.15	0.05

and the co-flow/burner assembly is mounted in a vertical wind tunnel with an exit cross section of 290 $\times$ 290 mm. The exit plane of the co-flow and nozzle is located 59.0 mm downstream of the exit plane of the wind tunnel. The co-flowing air velocity is fixed at 4.5 m/s, and the pilot unburnt velocity is fixed at 1.5 m/s with an equivalence ratio of 1.0 so that the main controlling parameters for the flames are the mass flow rates of liquid fuel and the bulk velocity of the carrier fluid. Stability limits are generated with respect to these parameters and are plotted elsewhere [37] for both acetone and ethanol fuels.

For each of the acetone and ethanol flames (referred to as AcF and EtF, respectively), six cases are chosen to study the effects of increasing the carrier velocity at a fixed liquid flow rate (referred to as sequence 1, 2, 5 and 7) or the effects of increasing the liquid fuel flow rate for a fixed carrier velocity (referred to as sequence 4, 3, and 1). To facilitate comparison, cases with the same numerical reference (such as AcF1 and EtF1) have similar mass flow rates for both carrier gas and liquid fuel. Three liquid fuel mass flow rates are used, namely 23.4, 45 and 75 g/min and these are referred as "Low", "Mid" and "High" respectively. Also, four carrier velocities of 24, 36, 48 and 60 m/s (corresponding respectively to carrier mass flow rates of 150, 225, 301 and 376 g/min) are employed as shown in Table 1. It should be noted that air has been used as carrier for all the cases studied in this chapter.

Joint LIF-OH-acetone is performed in acetone flames. The $Q_1(6)$ line of the (0,1) vibrational band of the $X^2\Pi$-$A^2\Sigma^+$ transition at 283.9 nm is used as a single laser source to simultaneously excite LIF-OH and acetone. The laser pulse energy was measured to be 4 mJ and the laser beam was focused into a 150 micron thick sheet using a 300 mm focal length cylindrical lens. Joint LIF-OH-CH_2O and Mie scattering was imaged in ethanol flames. The set-up for LIF-OH remains unchanged as described above for acetone. Formaldehyde (CH_2O) was excited in $4^1{}_0$ vibrational transition in the $\tilde{A}^1A_2$–$\tilde{X}^1A_1$ band near 353.17 nm and this same line was used to collect Mie scattering from ethanol droplets at 353.17 nm. LIF-OH-acetone and LIF-OH-CH_2O was collected on two intensified Flowmaster CCD cameras located on opposite sides of the flame. The LIF-OH signal is filtered using a high pass WG-295 filter and an interference filter centred at 310 nm with a 10 nm bandwidth while broadband LIF-acetone is filtered using a Schott glass filter GG-400 and low pass filter SPF-450. The broadband LIF-CH_2O signal is also filtered using a Schott glass filter GG-400 and low pass filter SPF-450. The droplet Mie signal is not intensified and passes through an interference filter centred at 355 nm with a bandwidth of 10 nm.

While it is known that interference from PAH molecules affects LIF-CH_2O particularly in fuel rich conditions, the question here is whether such effects are significant in the regions of heat release so that the product CH_2OxOH is corrupted by PAH. Simple laminar flame calculations using a detailed mechanism that includes some PAH species confirm that there is no overlap between the mass fraction profiles of the representative PAH molecules and the rate of heat release represented by the product of CH_2OxOH. Hence, even though PAH interferes with CH_2O in rich mixtures ($\xi > 0.1$), the region where heat release is significant as marked by CH_2OxOH, is generally free from such interferences

2.2 *Compositional Structure of Flames*

Figure 1 shows samples of representative simultaneous LIF-OH-acetone images measured at various axial stations in the five acetone flames listed in Table 1. The measurements are presented here at upstream locations of $x/D = 5$, 10 and 15, where x is the axial distance and D is the spray jet diameter. The trends shown by flames AcF4, AcF3 and AcF1 represent the effects of increasing fuel flow rate (at a fixed air carrier velocity) while the trends shown by flames AcF1, AcF5 and AcF7 represent the effects of increasing carrier velocity (at a fixed fuel loading). The top and middle frames of each image set represent LIF-OH and LIF-acetone respectively and the third frame contains the superimposed image of LIF-acetone-OH. The individual frame size is 29×15 mm. On the intensity color scale, black represents 0 and red maximum intensity, while white represents a saturated signal. LIF-acetone was collected on an intensified camera setup in order to obtain a high signal to noise ratio from the vapor and thus LIF intensity from the large liquid acetone droplets are saturated and are marked by white spots on the LIF image. The droplet size on the images is considerably larger than the actual droplet and this is due to intensifier blooming effect. Close to the jet exit plane, all the fuel is confined to the center of the jet and is shrouded by the pilot flame. Large amounts of fuel evaporate inside the burner and as the fuel loading increases the amount of acetone vapor also increases creating a non-combustible rich mixture of fuel and air exiting the nozzle. This is confirmed by the large values of Φ_{exit} reported for these flames in Table 1.

At $x/D = 5$, flame AcF4 shows broad OH fields wrapping around the outer edges of the acetone fuel almost without any separation. This is different from flames AcF3 and AcF1 which show thin smooth OH zones with clear separation between the fuel and the reaction zone (marked tentatively by the presence of OH). It is worth noting that increasing the carrier velocity in flames AcF5 and AcF7 reduces this separation again such that the structure for flames AcF4 and AcF7 are similar. Further

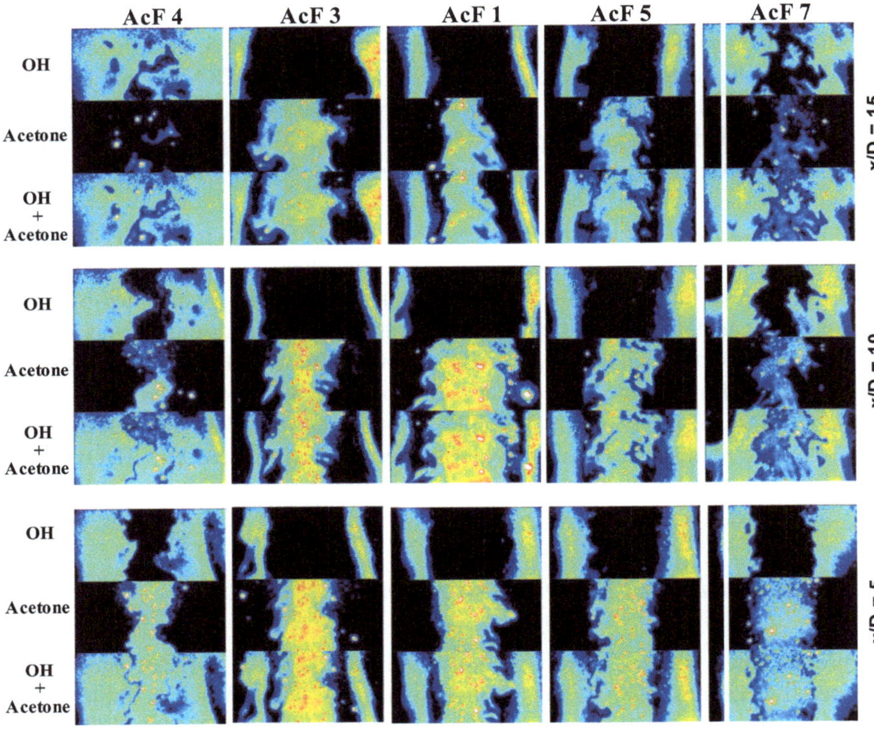

Fig. 1 Simultaneous and instantaneous LIF images of OH, acetone and acetone/OH image combined together (OH + Acetone) measured at x/D = 5, 10 and 15 in flames AcF4, AcF3, AcF1 AcF5 and AcF7

downstream at x/D = 10, the concentration of liquid acetone in the centre of the jet reduces significantly for flames AcF4 and AcF7 while the OH zones are even broader which is more akin to premixed flames rather than non-premixed particularly at such upstream locations. Interestingly, flames AcF3 and AcF1 show a totally opposite trend where the fuel zones are even wider and the OH fields are thinner and pushed further away from the jet centerline as expected for diffusion flames. At x/D = 15, the OH fields cover the entire image for flames AcF4 and AcF7 with scattered patches of OH even on the centerline wrapping around pockets of cold unreacted fuel that may contain droplets of acetone. Flames AcF3 and AcF1 exhibit very little change in the overall structure at this location and show, along with flame AcF5, thin OH zones surrounding a wide field of rich mixture of fuel and carrier air in the center of the jet.

The pattern observed in Fig. 1 for acetone flames is interesting and can be summarized as follows: at sufficiently low liquid fuel loading and carrier air velocity (AcF4), the gas emerging from the jet exit plane appears to be sufficiently premixed to form broad OH profiles and hence broad reaction zones surrounding the central fuel core which reduces in concentration further downstream leading to further broadening in the reaction zones. As the liquid fuel loading increases for the same carrier velocity (AcF3 and AcF1), a diffusion flame structure is gradually resumed with relatively thin outer reaction zones clearly separated from the central spray core. This structure persists from x/D = 5 to 15. With increasing the carrier velocity for the same fuel loading (as for AcF1, AcF5 and AcF7) the trend is reversed and a structure not unlike that of AcF4 is observed for flame AcF7 with AcF5 being a transitional flame from diffusion to premixed-like. It is worth noting that this picture is consistent with the values reported in Table 1 for Φ_{exit} where the richest mixture at the jet exit plane is obtained for flame AcF1 which takes a diffusion-like structure.

Representative, simultaneous LIF images of OH, formaldehyde (CH_2O), droplet Mie scattering and heat release, HR as determined by the product (LIF-OH × LIF-CH_2O) are presented in Fig. 2 for three axial locations in the five ethanol flames listed in Table 1. The trends shown by flames EtF4, EtF3 and EtF1 represent the effects of increasing fuel flow rate (at a fixed air carrier velocity) while the trends shown by flames EtF1, EtF5 and EtF7 represent the effects of increasing the carrier velocity (at a fixed fuel loading). The physical size of the individual image is 34 × 14 mm. The bottom image of each set (labeled HR + Droplet) shows the droplet positions with respect to heat release regions by superimposing the HR image on that of droplets (Mie scattering). The HR + Droplet images were regenerated, for clarity, to have only three color codes where yellow marks the HR region, pale blue marks the droplets and red marks the regions of overlap between HR and droplets. Mie scattering signals from liquid droplets are saturated in order to show small as well as large droplets. Some of the droplets are artificially large due to intensifier blooming effect. The spatial resolution of the droplet image is 25 microns so droplets smaller than 25 μm are not resolved.

At x/D = 5, flame EtF4 shows a relatively broad reaction zone as represented by the OH image which surrounds the LIF-CH_2O image. The heat release zone lies on the inner edge of this OH region as indicated by the images of HR which in turn surrounds the inner droplet as shown in the image of HR + Droplet. Increasing the droplet loading for the same carrier velocity (EtF3 and EtF1) leads to the appearance of pockets of OH on the centerline of the jet and these are separated from the outer OH profile which becomes thinner and diffusion like in flame EtF1. This double reaction zone structure is confirmed in the images of HR for flame EtF1 at x/D = 5. It is also interesting that the HR + Droplet images show some overlap between HR and the droplets which is not simple to interpret. One possible explanation is that droplets are much smaller than the width of the imaging laser sheets and hence become partly embedded within. Increasing the carrier velocity for the same droplet loading (cases EtF1, EtF5, EtF7) causes the otherwise double reaction zones to mesh together as seen at x/D = 5 in flame EtF5. This is further confirmed in flame EtF7 where a single, broad layer for heat release is obtained at all axial locations. The similarities between the structure of flames EtF4 and EtF7 are also noted here.

Further downstream at x/D = 10, the transition from a single broad region in flame EtF4 to a double reaction structure in flames EtF3 and EtF1 is now much clearer as reflected both in the OH as well as the HR images. Flame EtF1 shows thinner OH zones than those of flames EtF4 and EtF3. Increasing the carrier air velocity for

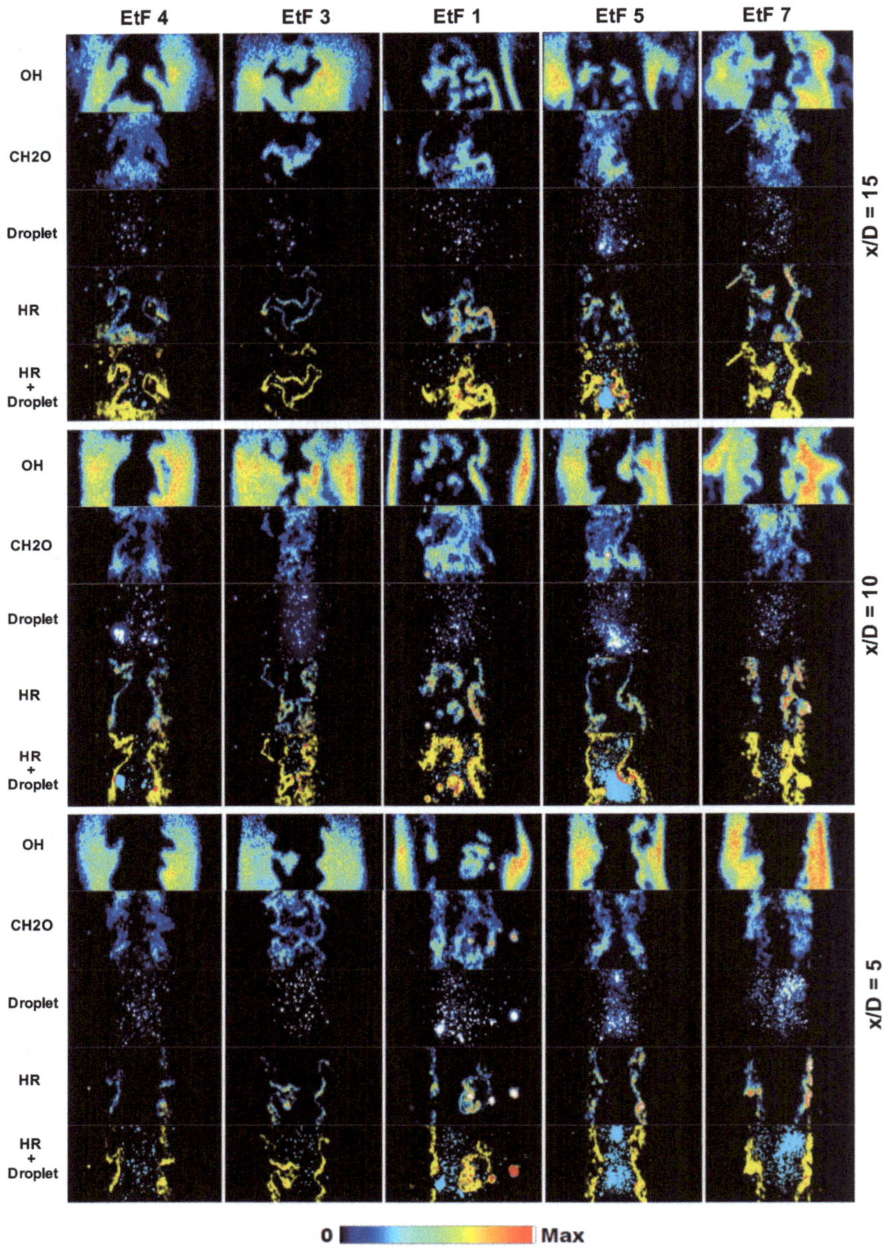

Fig. 2 Simultaneous images of LIF OH, LIF CH_2O, droplet Mie scattering Droplet), heat release ($HR = OH \times CH_2O$) and superimposed image of HR onto the droplet image (HR + Droplet) collected at x/D = 5, 10 and 15 in flames EtF4, EtF3, EtF1, EtF5 and EtF7

the same liquid loading as flame EtF1 (EtF5 and EtF7) broadens and contorts the OH profiles and the double reaction zones are brought closer together but not fully merged even in flame EtF7. It is not until x/D = 15 that these double reaction zones merge for flame EtF7 but not EtF5. It is also interesting that the transition from single broad OH profiles to a double reaction zone occurs also at x/D = 15 as the liquid fuel loading increases from flame EtF3 to EtF1.

Evaporation of ethanol droplets inside the burner is not as significant as acetone, this is evident from the values of Φ_{exit} shown in Table 1. The equivalence ratio at the jet exit plane increases as the fuel loading is increased to a maximum of 1.75 for EtF1 and then decreases with increasing carrier air velocity. Locally flammable mixtures may be formed at the jet exit plane of flame EtF1 and this is evident from the double reaction structure imaged here. Conversely, flames EtF4 and EtF7 have values of Φ_{exit} that are lean on average and hence double reactions in these flames are not expected. In most flames, the regions of heat release form on the periphery of the fuel vapor/droplet cloud but there are also instances where broad HR regions form in the center of the jet. The latter feature is more common in the lower velocity flames EtF1 and EtF3.

It is evident from these results that computing the compositional structure requires the correct representation of not only the chemical and physical properties of the fuel but also the interaction of these with the turbulence field. Droplet evaporation rate, droplet dispersion, turbulent mixing and chemical kinetics need to be modeled adequately to reproduce the differences noted between the acetone and ethanol cases discussed here. Such a challenge is likely to form the subject matter of future TCS workshops.

3 Auto-Ignition of Dilute Sprays

This section provides a summary of recent studies on the auto-ignition of dilute sprays in hot vitiated co-flows. A range of liquid fuels are tested and a discussion of the mechanism of auto-ignition is presented along with representative results.

3.1 The Burner

The Cabra [38] configuration which uses a hot co-flow of combustion products from lean premixed hydrogen flames is adopted here but modified to accommodate spray fuels as shown in Fig. 3. Extensive details about the burner design may be found elsewhere so only a brief description is given here. An ultrasonic nebulizer (SonoTek) with a 45° spray cone angle is placed just upstream of the central fuel tube and is surrounded by a co-flowing stream of carrier gas (either air or nitrogen). The nebulizer produces polydisperse sprays with a diameter range of approximately 1–100 μm and a Sauter mean diameter of approximately 40 μm. The droplets are carried by nitrogen

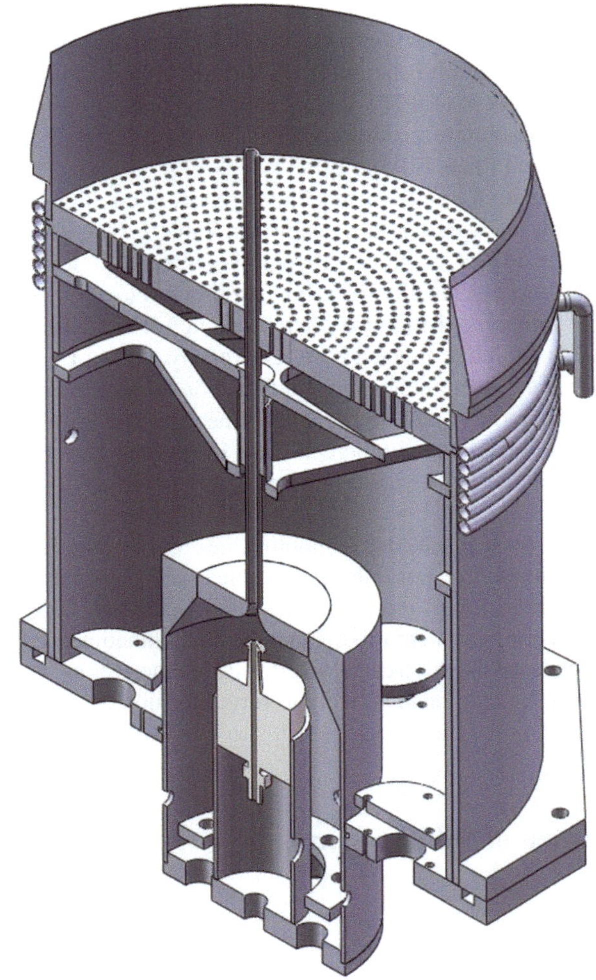

Fig. 3 Isometric sectional view of the auto-ignition burner

or air to the burner exit through a 196 mm long fuel tube with an inner diameter of 4.6 mm. The outer co-flow is 197 mm in diameter bringing hot vitiated products from lean premixed hydrogen/air flames stabilized on a perforated plate situated 68 mm upstream of the exit plane of the fuel tube.

The stability limits of the burner are characterized in terms of three main controlling parameters: (i) the temperature of the hot co-flow (T_c), (ii) the bulk jet velocity of the spray carrier (U_{jc}), and (iii) the mass ratio of fuel over carrier in the central jet (M_f/M_c) (also referred to as "fuel loading"). For a fixed value of T_c and U_{jc}, and with increasing the fuel loading, the flame lift-off height above the exit plane of the fuel pipe, L_h decreases and the shape of the flame changes as noted from the photographic

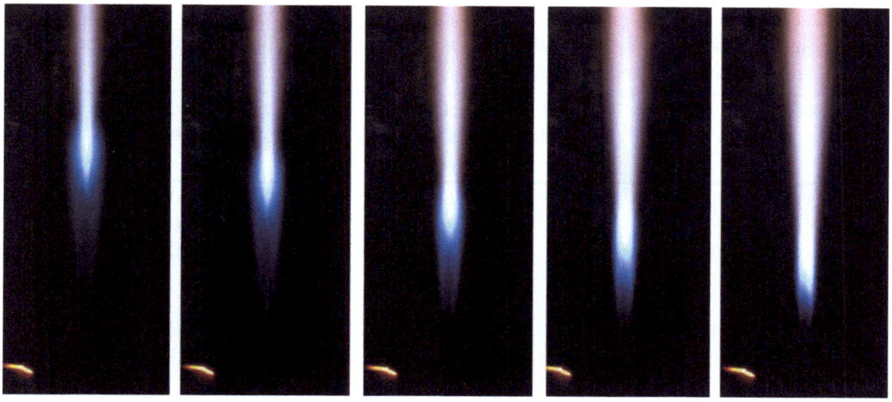

Fig. 4 Photographic images of auto-igniting ethanol flames at various fuel mass flow rates. The first three images on the LHS belong to type A, the fourth image is of type B and the image on the far right is type C

images shown in Fig. 4 for flames of ethanol fuel. Three qualitative flame shapes are identified. Type (A) features a conical yellow shape downstream of a broad light blue base. With increasing fuel flow rate, the tip of the intense cone-like region moves gradually upstream until such a point where the central fuel jet penetrates through the base of this cone-like structure forming two annular intense blue regions downstream of the light-blue flame. This annular structure characterizes type C flame and persists along much of its visible length. Type B is a transitional condition between A and C.

3.2 *Flow and Droplet Fields*

Five liquid fuels (methanol, ethanol, acetone, hexane and heptane) are studied at different co-flow temperatures, T_c and bulk carrier velocities, U_{jc} as listed in Table 2. The stability limits for these fuels were reported earlier [31], [39] and are not reproduced here. Two non-reacting cases of acetone and turpentine (Ac1A-NR and TURP) are also studied with turpentine representing a base case that has little or no evaporation. Methanol is investigated with both air and nitrogen as carrier and the effects of the carrier change were found to be minimal as reported in [31], [39]. For each reacting condition, three flames with different liquid loading (M_f/M_c) were selected for further investigation. These are referred to as cases A, B, C having different lift-off height and flame appearance. The lift-off height, L_h is defined here as the mean location of the flame base where light blue emissions are detected on still images taken with a normal digital camera.

Details on the velocity and droplet fields, as well as the auto-ignition characteristics of selected flames of ethanol and methanol are reported elsewhere [31],

Table 2 Liquid fuels investigated for auto-ignition listing some relevant properties for the selected conditions. Flames A, B and C are selected for each reactive case. Successive values tabulated for M_f/M_c are for cases A, B, and C, respectively

Designation	Fuel	Co-flow Temp. (K)	Fuel Loading (M_f/M_c)	Jet Carrier Velocity (m/s)	Jet Reynolds Number	Carrier Gas
TURP	Turpentine	N/A	0.260	75	23750	Air
Ac1A-NR	Acetone	N/A	0.212	75	23750	Air
Et1A-B-C	Ethanol	1380	0.253, 0.284, 0.316	75	23750	Air
Mt1A-B-C	Methanol	1380	0.228, 0.259, 0.294	75	23750	Air
Mt2A-B-C	Methanol	1430	0.228, 0.259, 0.294	75	23750	Air
Mt3A-B-C	Methanol	1430	0.233, 0.279, 0.323	60	19000	Air
Mt1AN-BN-CN	Methanol	1380	0.228, 0.259, 0.294	75	23400	Nitrogen
Mt2AN-BN-CN	Methanol	1430	0.228, 0.259, 0.294	75	23400	Nitrogen
Ac1A-B-C	Acetone	1380	0.212, 0.247, 0.281	75	23750	Air
Hx1A-B-C	Hexane	1320	0.165, 0.234, 0.270	75	23750	Air
Hp1A-B-C	Heptane	1320	0.203, 0.244, 0.286	75	23750	Air

[39]. A comparative study of the droplet dynamics between non-reacting and reacting spray jets of acetone is also available [40], [41]. The entire set of results is documented in the PhD thesis of O'Loughlin [41] and may be made accessible on request. Mean velocities and rms fluctuations as well as profiles of volume flux, droplet size distributions and number densities are measured at a range of axial locations in the flames. To facilitate implementation for computations, a detailed account of the boundary conditions is also documented for each condition [41], including the selected non-reactive cases.

Figure 5 shows sample results for flame Et1B which is representative of ethanol flames with $T_c = 1380K$, $U_{jc} = 75$ m/s and $M_f/M_c = 0.284$ (see Table 2). Radial profiles of the mean axial velocity, the axial velocity rms, and axial volume flux of liquid as well as the droplet number densities are shown for $x/D = 0.4$, 10, 20 and 30. Results are conditioned on the droplet size binned into five diameter ranges, 10

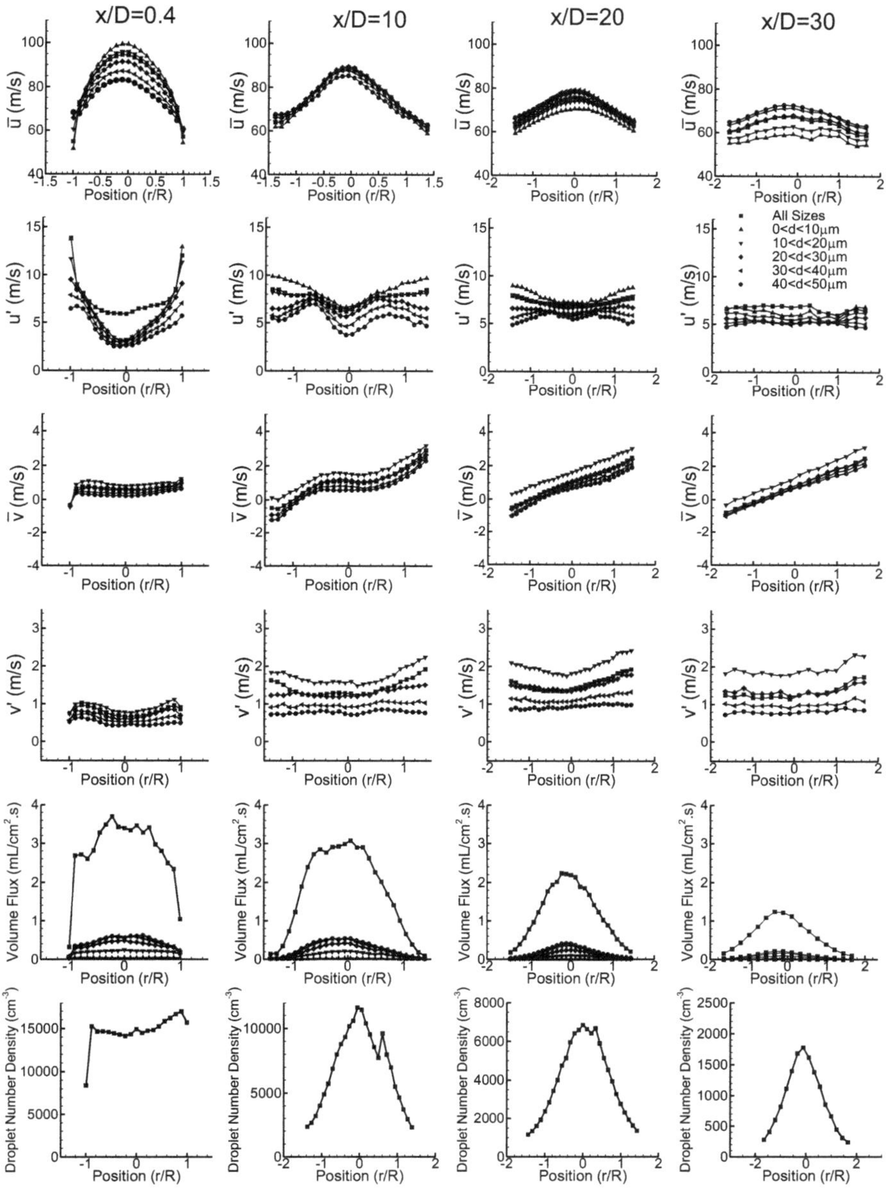

Fig. 5 Profiles of mean axial velocity, u, its rms of fluctuations, u′, mean radial velocity, v, its rms of fluctuations, v′, axial volume flux and droplet number densities measured at three axial locations (x/D = 0.4, 10, 20 and 30) in flame Et1B

microns wide. The condition at x/D = 0.4 represents the boundary condition while the others show the evolution of the droplets with downstream distance. The following observations are made from this plot as well as from others that are available elsewhere [31], [39]:

1. Similar to the behavior of dilute spray flames in ambient flows, large droplets exit the jet with negative slip velocities which then undergo a reversal to positive values around x/D = 15.
2. While the axial rms velocity fluctuations for small droplets remain higher than those of large ones, the trends in the profiles are different at upstream axial locations of x/D < 30 and particularly away from the centerline. For large droplets, the peak rms remains somewhat uniform around 6 m/s while small droplets show a significant increase to about 10 m/s at outer radial locations and this may be caused by selective droplet dispersion. It is noted that the peak rms fluctuations for small droplets decrease with axial distance and the profiles for all droplet sizes tend to flatten and draw closer around x/D = 30.
3. The radial mean velocity plots confirm that the centerline radial velocity is zero while profiles of the radial rms velocity consistently show that larger droplets have significantly lower radial rms (up to three times) than the smallest droplets.
4. While large droplets dominate the volume flux, profiles of droplet number densities show that the bulk of the droplets are contained in the two smallest diameter bins $0 < d < 10\ \mu m$ and $10 < d < 20\ \mu m$. Further downstream, the number density decays significantly due to the preferential evaporation of smaller droplets. This effect is most obvious near the periphery of the spray where the temperatures are highest and therefore evaporation is fastest.

3.3 *Reactive Fields*

The images presented in Fig. 4 highlight the fact that these flames may be nominally split into two regions: (i) an auto-ignition zone just downstream of the jet exit plane where the radical pool builds up and (ii) a flaming zone further downstream marked by the presence of a luminous, yellow flame with significant heat release. High-speed (5 kHz) imaging of laser induced fluorescence from OH (HS-LIF-OH) were found to be extremely informative in identifying modes of ignition. Similarly, low-speed (10 Hz) joint imaging of OH and CH_2O yields a useful measure of heat release through the product [OH] × [CH_2O] and these were reported earlier for selected flames of methanol [39].

High-speed images in the auto-ignition zone reveal the presence of kernels which seem to be responsible for initiating heat release further downstream. This is very consistent with the auto-ignition mode reported earlier for gaseous flames [42], [43]. Figure 6 shows a sample illustration of two LIF-OH sequences measured in methanol flame Mt1C. The top row represents a successful event where a kernel continues to grow while the bottom sequence shows a failed event where an initiated kernel subsequently extinguishes. The occurrence of such kernels is noted for all fuels studied

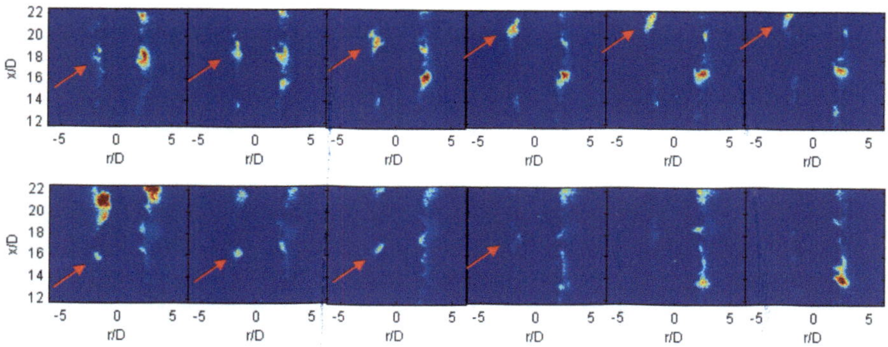

Fig. 6 High speed LIF-OH images in methanol flame Mt1C showing a successful kernel (top) and a failed kernel (bottom). The consecutive images are correlated in time with an interval of 200 μs separating each image

here and is therefore not a fuel specific phenomenon. Kernels may fail due to various reasons that may include a local increase in the scalar dissipation rate or quenching of ignition radicals by droplets or cold fluid transported from other locations in the flow. While it may also be argued that such failures are artifacts of "out-of-plane motion", this is highly unlikely considering the frequency of occurrence of these events. Experiments are currently being conducted to confirm this hypothesis.

While the mode of auto-ignition at the leading edge of the flames is somewhat common, the spatial structure of reaction zones further downstream seems to vary with the fuel used. Figure 7 shows high speed LIF-OH images of three sets of flames Et1A-C, Mt1A-C and Hx1A-C for the range from x/D = 2 to x/D = 42. Note that images of Hx1A-C reveal the droplets that have not been totally filtered out as well as OH. The initial OH formation occurs, as expected, on the periphery of the jet and does not extend to the centerline. Further downstream, however, significant occurrences of OH forming on the centerline are observed in flames Et1A and Et1B at x/D = 32–42. This suggests that premixing of the fuel/air mixture is occurring and there are sufficiently few droplets remaining in this area so that combustion on the centerline can occur. Flames Mt1A and Hx1A show that even at x/D = 32–42, the OH profiles remain well separated with droplets being present along the centerline as is evident from the images of Hx1A. An increase in the fuel mass loading causes a significant change in the flame structure. While the OH separation increases for flame Hx1C, the reverse is true for methanol with flame Mt1C showing OH present on the centerline. Also, the tendency for burning to occur on the centerline is significantly reduced for the ethanol flame with the increased fuel loading. It is worth noting that the methanol flames show very weak OH zones only on the periphery of the jet and these tend to occur further downstream by comparison to ethanol and hexane.

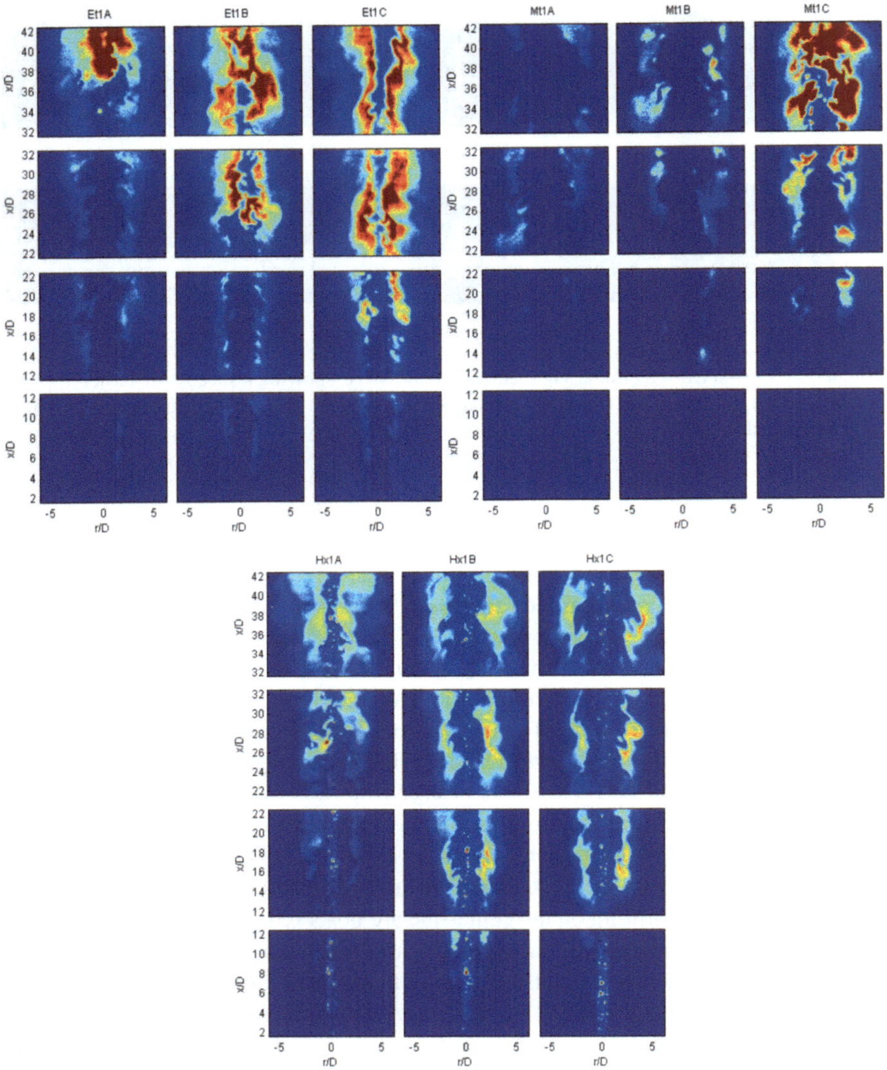

Fig. 7 High speed LIF-OH images of a set of three ethanol flames (top left), three methanol flames (top right) and three hexane flames (bottom) at a range of axial locations from x/D = 2 to x/D = 42. The co-flow temperature is 1380 K in all cases

4 Dense Sprays and Turbulence Effects on Secondary Atomization

As predictive capabilities for the structure of dilute spray jets and flames improve, a gradual shift of focus to denser sprays will occur. This will bring significant additional complexities associated with both the primary and secondary atomization regions as well as the significant interactions within the liquid elements and between the liquid and the gas phases. This section addresses only a single issue highlighting the effects of turbulence on atomization. It should also be noted that the definition of dense sprays varies and may be given in terms of liquid loading or optical density of the fluid exceeding a certain threshold.

The primary break-up region of a spray jet is probably the least understood and the most complex in dense sprays. It is dominated by instabilities that develop at the liquid-gas interfaces leading to disintegration processes that form droplets, fluid membranes or ligaments [18], [22] while also being a function of internal liquid jet turbulence, bubble formation, and gas phase turbulence. Secondary atomization is characterized by the droplets or ligaments interacting with the surrounding dispersed phase to further break-up into smaller fragments. This phase is much better understood than the primary break-up region [24]. The Weber and Ohnesorge numbers remain the main controlling parameters for transitioning into the various secondary break-up modes which are referred to (sequentially with increasing Weber number for a fixed Oh number) as vibrational, bag, multi-mode, sheet thinning, and catastrophic modes of break-up.

The main dimensionless numbers used to delineate the various modes of transition from droplets into 'children' droplets are the droplet Weber, Ohnesorge and Reynolds numbers We_d, Oh_d, Re_d defined here as follows:

$$We_d = \rho_g d(U - U_g)^2/\sigma_L \qquad Oh_d, = \mu_L/(\rho_g d\ \sigma_L)^{0.5} \qquad Re_d = \rho_g d|U - U_g|/\mu_g$$

where d is the droplet diameter, U and U_g are the droplet and gas velocities, μ is the dynamic viscosity, σ is the surface tension coefficient, ρ is density and subscripts g and L refer to the gas and liquid phases, respectively. Details on other relevant non-dimensional scalings in secondary atomization have been fully provided elsewhere [24].

The influence of turbulence on the modes of secondary break-up is a topic that has not received considerable attention despite the dominance of turbulent flows in sprays. The topic was originally explored by Kolmogorov [44] and subsequent experimentation was carried out by others [25], [45]. Nevertheless, further measurements using modern diagnostic techniques are required in order to fully understand the contribution of velocity fluctuations toward secondary atomization. The section addresses such issues using two configurations: (i) a simple pipe flow carrying droplets at various Reynolds number and turbulence levels and (ii) a dedicated air-blast atomizer where the turbulence intensity varies with the air-carrier loading and liquid fuels of different properties are subsequently tested. The air-blast atomizer has also been designed in order to allow for a systematic study of atomizing dense sprays in both non-reacting and reacting environments.

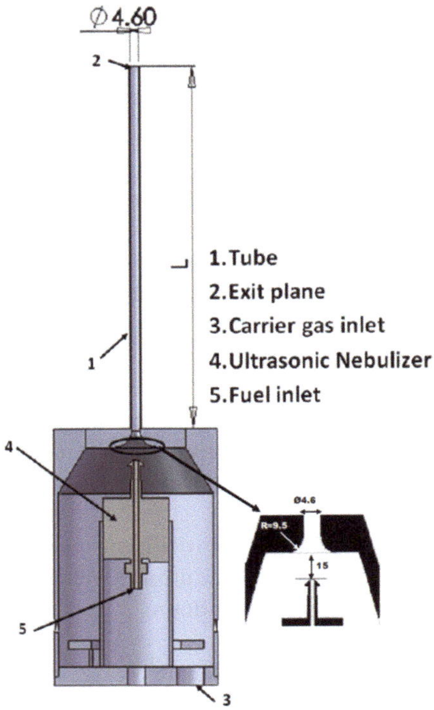

Fig. 8 Schematic of the dilute spray pipe

4.1 *The Spray Pipe*

This configuration, shown schematically in Fig. 8, uses the same method of atomization as the auto-ignition burner described earlier. An ultrasonic nebulizer is placed upstream of the inlet of a pipe with an internal diameter of D = 4.6 mm (see inset of Fig. 8). Three pipe lengths, L are used, namely with $L_1 = 22$ mm, $L_2 = 101$ mm and $L_3 = 196$ mm. The distance from the nebulizer tip and fuel tube entry point was chosen based on previous experiments [15]. Air is used as the carrier and, for each pipe length, its velocity and Reynolds number is increased to induce additional turbulence in the flow. A standard LDV/PDA system is used to perform measurements of velocity, turbulence and droplet size distribution at the exit plane of the pipe for each flow configuration.

Figure 9 shows evolution of the centreline Sauter mean diameter (D_{32}) measured at the exit plane of the various tube lengths over a range of Reynolds numbers. Two points are worth noting: (i) the mean droplet diameter, D_{32} decreases consistently with increasing Reynolds number and (ii) while little difference exists between the short and intermediate tubes (L = 4.7D and 22D), a significant drop in D_{32} is noted when the full tube length (L = 43D) is used.

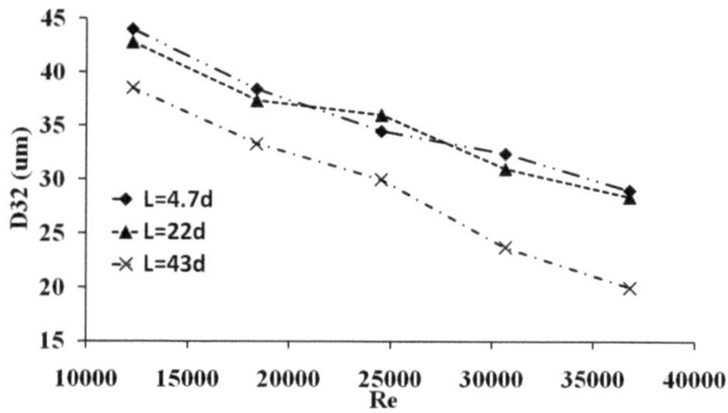

Fig. 9 Measured D_{32} on the centreline exit of the pipe for various pipe lengths and Reynolds numbers

This suggests that modes of secondary break-up are also affected by turbulence. Droplets are not simply fragmented at the inlet but continue to undergo further atomization within the pipe, particularly at high turbulence levels. The reader should note that an increase in velocity will not only increase the Reynolds number which will result in higher turbulence levels but it will also increase the Weber number resulting in increased shear at the droplet-air interface, which would heavily contribute to atomization. However, in this spray, droplets do not atomize solely due to higher Weber numbers and this claim can be substantiated by calculating the instantaneous Weber number for these droplets as shown in Fig. 10 which are of the order $We_d < 10$, thereby indicating vibrational deformation modes only [24]. If conventional atomization were occurring, then a Weber number of approximately $We_d > 11$ would indicate a bag-breakup regime with larger Weber numbers indicating multi-modal break-up and shear break-up regimes.

Going from the short to the long tube (RHS to LHS of Fig. 10) generally yields a decrease in droplet Weber number. However, this change also coincides with a decrease in droplet size (see Fig. 9), suggesting that it is not a conventional droplet atomization regime which is resulting in secondary atomization. The Weber numbers of the droplets, while not enough to create shear break-up, are sufficiently high to create oscillatory deformation. The data shown here suggest that turbulence, rather than a conventional break-up, is causing the secondary atomization observed in the spray, and the impact of the turbulent fluctuations may be enhanced due to the fact that the flow is acting on deformed droplets.

For droplets to be able to atomize through the action of velocity fluctuations they must be greater than the Kolmogorov length scale, $\eta = (\nu_g^3/\varepsilon)^{1/4}$, where ν_g is the kinematic gas viscosity and ε is the dissipation. A higher Reynolds number for a particular spray will result in (i) an increase in the local rms fluctuating velocity which will partly define the smallest length-scale in the flow, and (ii) a reduction in the droplet size. Therefore, it is not immediately obvious if the droplets at a particular location will be larger or smaller than η. However, assuming that the droplets are breaking up due to the turbulent fluctuations, then by definition, they must always be greater than the Kolmogorov length.

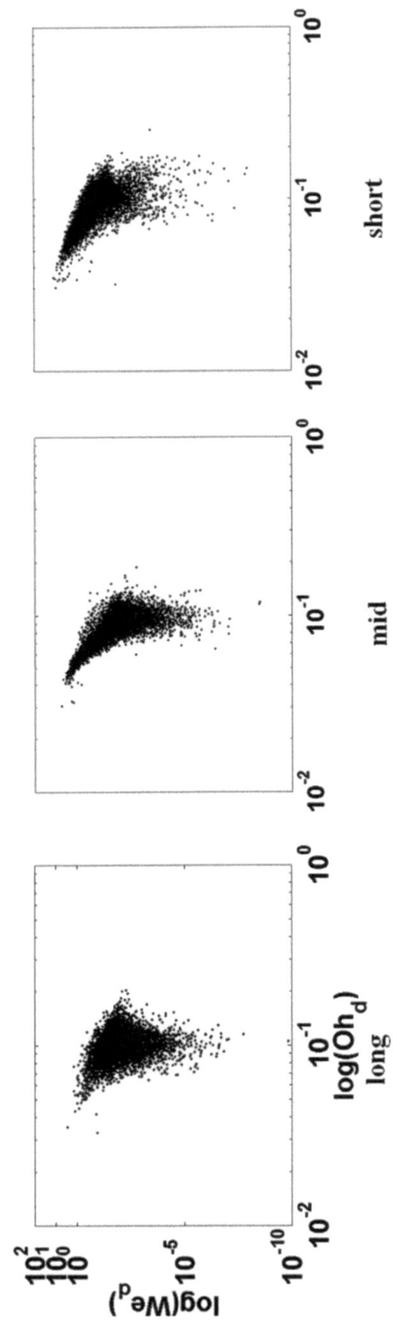

Fig. 10 Log-log scatter plots of Weber number vs. Ohnesorge number for the long, mid and short tube cases for an air jet velocity $U_g = 120$ m/s, and a fuel mass flow-rate $M_f = 28.7$ g/min

The Kolmogorov length scale is calculated with the dissipation set to $\varepsilon = {u'_g}^3/\Lambda$ where u'_g is the rms axial velocity of the gas phase and Λ is the integral length-scale, estimated as $\Lambda = 0.65R_h$ where the half-radius $R_h = 0.0845X_L$ and X_L is the axial location downstream. Figure 11 shows the Sauter mean diameter versus r/D for increasing Reynolds numbers (A2 to A6) all conditioned on the d = 0–10 μm droplet range. The trend of D32/η generally follows the trend of the axial rms velocity, given the relationship used between the Kolmogorov length and u′. Clearly, in the 0–10 μm range, none of the cases from A2-A5 are able to atomize at the spray centreline under the influence of turbulent fluctuations, but theoretically, case A6 may be prone to turbulence enhanced atomization. However, even for case A6, across the full radial profile, the droplet sizes are only 1.5–2 times greater than the Kolmogorov length, indicating that it is generally unlikely that any droplets so small would atomize. In the 20–30 μm range of Fig. 11, it may be seen that the D32 is now at least twice the Kolmogorov length scale in the centre of the spray and can increase to as high as one order of magnitude larger in the shear layer.

These results suggest that droplets located at r/D = 0.4 would be more prone to break-up due to turbulence. However, it is difficult to substantiate this claim from the measurements given that droplets detected at r/D = 0.4 may also have migrated from the centre to due radial dispersion. Nevertheless, the results do show a consistent decrease in droplet size when moving from the shorter to longer tube lengths and this is clearly more severe for higher Reynolds numbers. Imaging of these sprays or similar sprays under the influence of high turbulence intensity using a high magnification lens would provide further information regarding the physics of the turbulent atomization in these systems. Such experiments are being performed.

4.2 Multiple Stage Atomizer

Isometric views of a multi-stage atomizer are shown in Fig. 12. While only the air-blast ports (port 2) are used here, the atomizer has the flexibility to operate in an effervescent mode (zone 1) as well as with the assistance of swirl (port 3). The impact of those stages on the atomization of the jet has been examined elsewhere [46]. In the experiments reported here, only co-axial air supplied through port 2 is utilised. The liquid nozzle diameter D_l fixed to part 1 is kept at a constant 0.5 mm for all experiments and the airblast nozzle diameter D of part 2 is fixed at 10 mm. The liquid injection nozzle (part 1) is located upstream of the air nozzle for a number of design reasons discussed elsewhere [46].

The effects of turbulence and higher jet velocities on the atomization are investigated. While results are shown here for ethanol only, a range of biodiesels with different physical as well as chemical properties have been studied and reported elsewhere [47]. In addition to measurements of velocity and droplet fields using the LDV/PDA systems, microscopic imaging at high speed is employed using a 10 kHz Nd-YAG laser as the illumination source in order to resolve the dynamics of the atomization process. On the detection side, a high-speed CMOS camera (LaVision) coupled to a long distance microscope objective lens (QUESTAR, QM-100) is used to image a 2.6 × 2.6 mm area of the flow with a resolution of approximately 3.3 microns. Further details about the experimental set-up and image processing procedure may be found elsewhere [47].

Fig. 11 Sauter mean diameter (D32) conditioned on droplet sizes d = 0–10 μm (left), d = 20–30 μm (mid) and d = 40–50 μm (right), normalized by the Kolmogorov length η and plotted vs. r/D as a function of Reynolds number for the mid-length tube where A2- > A6 indicates an increase in Reynolds number from 12300 to 36800

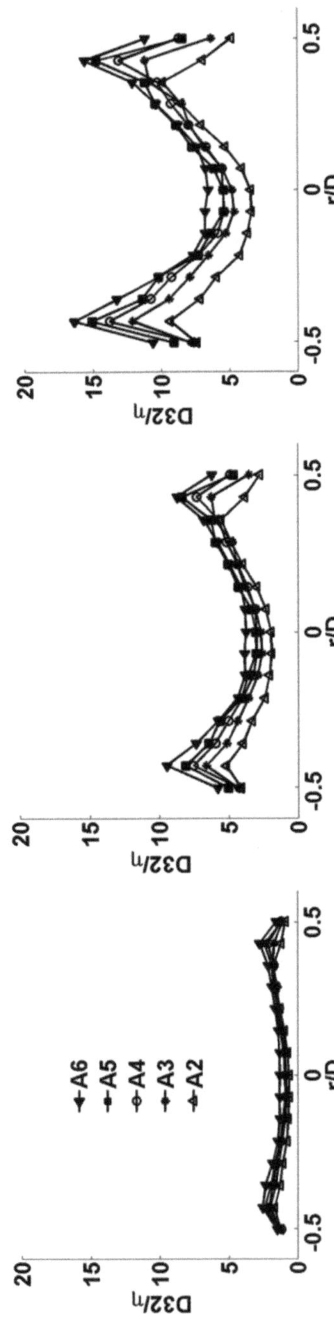

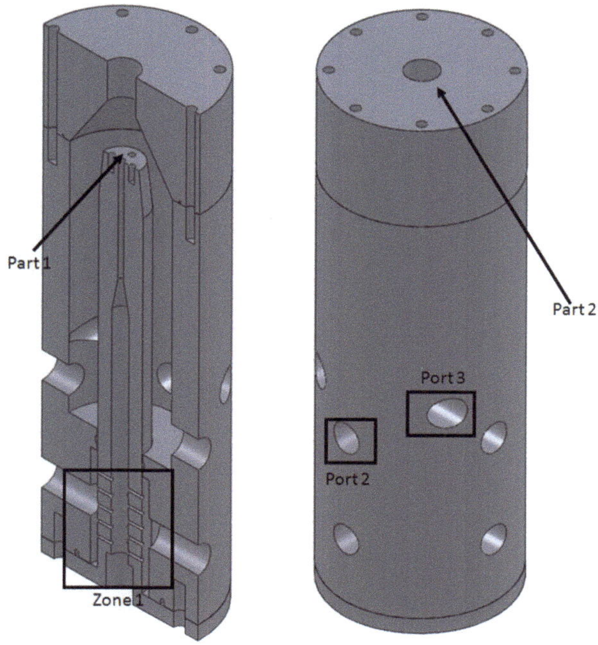

Fig. 12 Schematic of atomizer where part 1 is the liquid nozzle, part 2 is the air exit nozzle, zone 1 shows the effervescent stage, port 2 shows one of the four air-blast inlet ports and port 3 shows on of the four swirl inlet ports. Figure appears in [47].

Figure 13 shows radial profiles of mean axial droplet velocity measured in three ethanol sprays referred to as EA5 ($U_{jexit} = 64$ m/s, $m_l = 0.04$ kg/min), EA6 ($U_{jexit} = 64$ m/s, $m_l = 0.07$ kg/min) and EA9 ($U_{jexit} = 74$ m/s and $m_l = 0.07$ kg/min). The liquid loading increases from EA5 to EA6 while the Reynolds number increases from EA6 to EA9. For all of the cases shown, a clear decrease in the axial mean velocity is noted at the exit plane when moving from outer radial locations towards $r/D = 0$ and this agrees with literature [48–49]. The 'dip' at the exit plane ($x/D = 0$) exists because the liquid jet travels at a lower velocity to that of the surrounding coflowing airblast velocity, causing the full range of droplets in the central core region to move slower. This phenomenon is particularly clear in the high liquid flow-rate cases (EA6 and EA9) of Fig. 13. Once the atomization process is complete, the decay in the axial mean velocity with downstream location is physically identical to that of the momentum decay in the dilute sprays examined earlier in this chapter.

Figure 14 shows radial profiles of D32 measured at various axial locations in ethanol sprays. Results are presented for cases EA5, EA6 and EA9 showing the effect of Reynolds number for fixed mass loading (EA6 to EA9) and the effect of mass loading for fixed Reynolds number (EA5 to EA6). Cases EA9 and EA6 of Fig. 14 resemble a radial droplet size distribution of a conventional coaxial air-blast atomizer where the liquid injection location is at the same location as the air nozzle

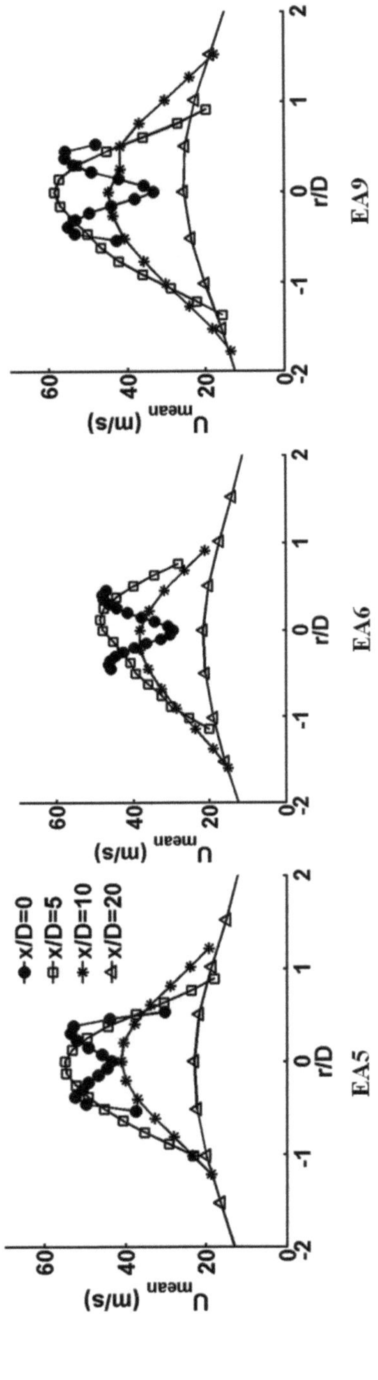

Fig. 13 Radial profiles of unconditional mean axial velocities measured in ethanol sprays at various downstream positions with $U_{jexit} = 64$ m/s and $m_l = 0.04$ kg/min (EA5), $U_{jexit} = 64$ m/s and $m_l = 0.07$ kg/min (EA6) and $U_{jexit} = 74$ m/s and $m_l = 0.07$ kg/min (EA9). Figure appears in [47]

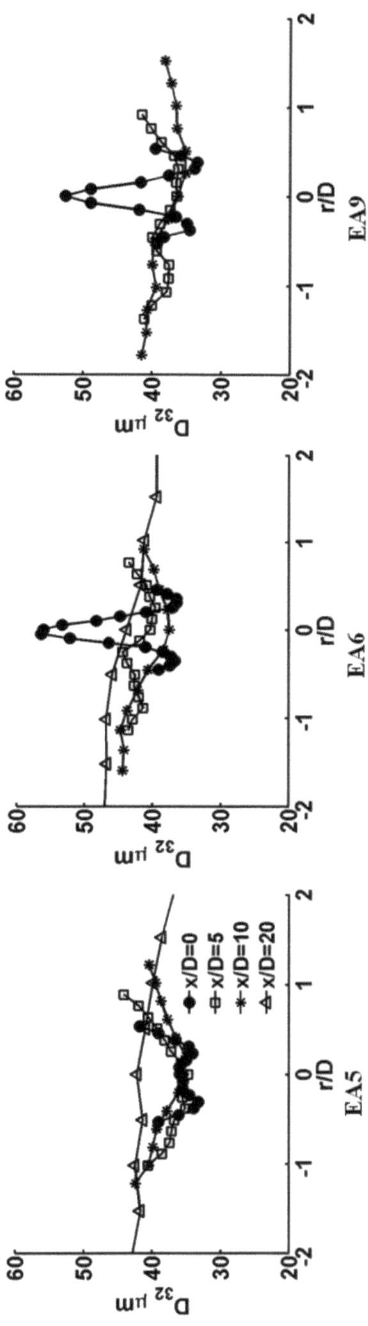

Fig. 14 Sauter mean diameter vs. r/D for ethanol for various downstream positions with $U_{jexit} = 64$ m/s and $m_l = 0.04$ kg/min (EA5), $U_{jexit} = 64$ m/s and $m_l = 0.07$ kg/min (EA6) and $U_{jexit} = 74$ m/s and $m_l = 0.07$ kg/min (EA9). Figure appears in [47]

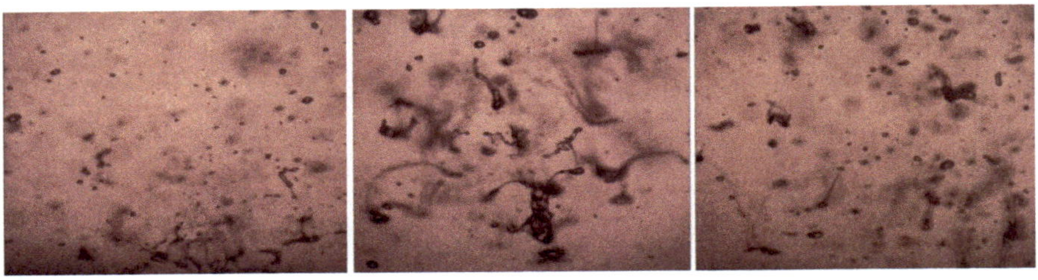

Fig. 15 Snapshots showing typical atomization events at the exit plane for three cases of ethanol, for a field of view of 2.6 × 2.6 mm with a spatial resolution ∼ 3.3 μm

exit plane. This is due to the much larger droplets present in the middle of the spray when compared to outer radial locations, and for the same reason as with the 'dip' in the axial mean velocity in the centreline, this occurs due to a lack of inter-phase mixing. This is unlike case EA5 of Fig. 14 where significant atomization occurs upstream of the exit plane therefore yielding an overall smaller droplet size in the core of the spray.

Figure 15 shows typical instantaneous shadowgraph images for sprays of ethanol for the same cases shown in Figs. 13 and 14, namely EA5, EA6 and EA9. The images were all taken at the exit plane, revealing how much information a PDA measurement would exclude in that area. Through the images one can clearly distinguish between individual droplets and the unbroken portions of liquid such as ligaments and highly deformed droplets.

As the liquid loading increases for a given Reynolds number (EA5 to EA6), a generally larger population of ligaments appears, given that more energy is required from the gas phase to atomize the same amount of liquid. An increase in the Reynolds number for a fixed liquid loading (EA6 to EA9) results in an improved degree of atomization which is expected. For the representative image of case EA9 shown in Fig. 15, this improved atomization manifests as a transformation from a predominant population of short and long ligaments to an increased population of slightly deformed droplets as detailed in [47]. Certain quantitative measures can be extracted from this region to give information on the density of the spray while also the distribution of liquid shapes, ranging from longer ligaments to shorter ligaments, deformed droplets, spherical droplets and large unbroken liquid volumes as fully described in [47]. Automating such a detection system will allow for a detailed set of quantitative data to assist with boundary condition implementation in computational models for denser sprays where PDA is ineffective as a measurement technique close to the exit plane.

5 Concluding Remarks

The structure of reaction zones in turbulent dilute sprays flames is complex and depends on many factors including the physical properties of the fuel (which affect its evaporation characteristics) as well as on the mixing patterns and the chemical kinetics. The latter is particularly relevant with respect to auto-ignition where the formation of radical pools and hence ignition kernels are critical for the initiation of heat release and hence flaming combustion. A range of liquid fuels tested in a hot, vitiated co-flow show similar modes of auto-ignition but different downstream zones of flaming characteristics.

The atomization region, while the most difficult to probe both in terms of modeling as well as diagnostics, remains critically important in shaping the remainder of the sprays. Results reported here demonstrate that turbulence affects the secondary atomization zone and hence adds another level of complexity to the calculations. The provision of improved data collected in well-defined model problems such as those discussed in this Chapter as well as the remainder of this Book will form an essential building block for advancing capabilities to models such complex flows.

Acknowledgement This work is supported by a grant from the Australian Research Council.

References

1. Nigam, P.S., Singh, A., "Production of liquid biofuels from renewable sources", Prog. Energy Combust. Sci. 37(2011) 52–68.
2. Demirbas, A., "Progress and recent trends in biofuels", Prog. Energy Combust. Sci. 33(2007) 1–18.
3. Drapcho, C.M., Nhuan, N.P., Walker, T.H., "Biofuels engineering process technology", McGraw Hill, 2008.
4. Alleman TL, McCormick RL., "Fischer–Tropsch diesel fuels—properties and exhaust emissions: a literature review" SAE paper 2003–01-0763, 2003.
5. Schaberg PW, Botha J, Schnell M, Herrmann HO, Keppeler S, Friess W., "HSDI diesel engine optimisation for GTL Diesel Fuel" SAE paper 2007-01-0027, 2007.
6. Agarwal, A.K., "Biofuels (alcohols and biodiesel) applications as fuels for internal combustion engines", Prog. Energy Combust. Sci. 33(2007) 233–271.
7. Lapuerta, M., Armas, O., Rodriguez-Fernandez, J., "Effect of biodiesel fuels on diesel engine emissions", Prog. Energy Combust. Sci. 34(2008)198–223.
8. Sun, J., Caton, J.A., Jacobs, T.J., "Oxides of nitrogen emission from biodiesel-fuelled engines", Prog. Energy Combust. Sci. 36(2010) 677–695.
9. Benjumea, P., Agudelo, J.R., Agudelo, A.F., "Effect of the degree of unsaturation of biodiesel fuels on engine performance, combustion characteristics and emission", Energy & Fuels 25(2011) 77–85.
10. Schonborn, A., Ladommatos, N., Williams, J., Allan, R., Rogerson, J., "The influence of molecular structure of fatty acid monolakyl esters on diesel combustion", Combust. Flame 156(2009) 1396–1412.
11. Wang, Y.L., Feng, Q., Egolfopoulos, F.N., Tsotsis, T.T., "Studies of C4 and C_{10} methyl ester flames", Combust. Flame 158 (2011) 1507–1519.
12. Gail, S., Sarathy, S.M., Thomsom, M.J., Dievart, P., Dagaut, P., "Experimental and chemical kinetic modeling study of small methyl esters oxidation: Methyl (E)-2-butenoate and methyl butanoate", Combust. Flame 155 (2008) 635–650.
13. Dooley, S., Curran, H.J., Simmie, J.M., "Autoignition measurements and a validated kinetic model for the biodiesel surrogate, methyl butanoate", Combust. Flame 153(2008) 2–32.
14. Farooq, A., Davidson, D.F., Hanson, R.K., Hyunh, L.K., Violi, A., "An experimental and computational study of methyl ester decomposition pathways using shock tubes", Proc. Combust. Inst. 32(2009) 247–253.

15. Herbinet, O., Pitz, W.J., Westbrook, C.K., "Detailed chemical kinetic mechanism for the oxidation of biodiesel fuels blend surrogate", Combust. Flame 157(2010) 893–908.
16. Ji, C., Dames, E., Wang, Y.L., Wang, H., Egolfopoulos, F.N., "Propagation and extinction of premixed C5-C_{12} n-alkane flames", Combust. Flame 157 (2010) 277–287.
17. Liu, N., Ji, C., Egolfopoulos, F.N., "Ignition of non-premixed C3-C_{12} n-alkane flames", Combust. Flame 159 (2012) 465–475.
18. Faeth, G., Hsiang, L., Wu, P., "Structure and breakup properties of sprays", International Journal of Multiphase Flow 21(1995) 99–127.
19. A. Lefebvre, "Atomization and Sprays", Taylor and Francis, 1989.
20. http://www.ercoftac.org/special_interest_groups/28_reactive_flows/upcoming_events/3rd_international_workshop_on_the_turbulent_combustion_of_sprays/.
21. http://www.sandia.gov/ecn/ECNworkshop.php.
22. Dumouchel, C., "On the experimental investigation on primary atomization of liquid streams", Experiments in Fluids 45(2008) 371–422.
23. Lasheras, J., Hopfinger, E., "Liquid jet instability and atomization in a coaxial gas stream", Annual Review of Fluid Mechanics 32(2000) 275–308.
24. Guildenbecher, P.D., Lopez-Rivera, C., Sojka, P., "Secondary atomization", Experiments in Fluids 46(2009) 371–402.
25. Hinze, J., "Fundamentals of the hydrodynamic mechanism of splitting in dispersion processes", A.I.Ch.E. Journal 1(1955) 289–295.
26. Linne, M.A., Paciaroni, M., Berrocal, E., and Sedarsky, D., "Ballistic imaging of liquid breakup processes in dense sprays", Proceedings of the Combustion Institute, 32(2009) 2147–2161.
27. Linne M.A., Sedarsky D., Meyer T., Gord J., Carter C., "Ballistic imaging of the flow in the interior of the near-field of an effervescent spray", Exp Fluids 49(2010) 911–923.
28. Kastengren A., Powell C.F., Liu Z., Wang J., "Time resolved, three dimensional mass distribution of diesel sprays measured with X-ray radiography", SAE technical paper series, paper no. 2009–01-0840, 2009.
29. Balewski, B., Heine, B., Tropea, C., "Experimental investigation of the correlation between nozzle flow and spray using laser doppler velocimeter, phase doppler system, high-speed photography, and X-ray radiography", Atomization and Sprays 20(2010) 57–70.
30. Wang, Y.J., Im, K.S., Fezzaa, K., Lee, W.K., Wang, J., Micheli, P., Laub, C., "Quantitative X-ray phase contrast imaging of air-assisted water sprays with high Weber numbers", Appl Phys Lett 89(2006):1–3.
31. O'Loughlin, W., and Masri, A.R., "A New Burner for Studying Auto-Ignition in Turbulent Dilute Sprays", Combust. Flame 158(2011) 1577–1590.
32. Markides, C.N., Mastorakos, E., "An experimental study of hydrogen autoignition in a turbulent co-flow of heated air", Proc. Combust. Inst. 30(2005) 883–891.
33. Neophytou, A., Mastorakos, E., Cant, R.S., "The internal structure of igniting turbulent sprays as revealed by complex chemistry DNS", Combust. Flame 159(2012) 641–664.
34. Neophytou, A., Mastorakos, E., Cant, R.S., "Complex Chemistry Simulations of Spark Ignition in Turbulent Sprays", Proc. Combust. Inst. 33(2011) 2135–2142.
35. Wang, Y., Rutland, C.J., "Direct numerical simulation of ignition in turbulent *n*-heptane liquid-fuel spray jets", Combust Flame 149(2007) 353–365.
36. Masri, A.R., and Gounder, J.D., "Turbulent Spray Flames of Acetone and Ethanol Approaching Extinction", Combust. Sci. Technol. 182(2010) 702–715.
37. Gounder, J.D., Kourmatzis, A., and Masri, A.R., "Turbulent Piloted Dilute Spray Flames: Flow Fields and Droplet Dynamics", Combustion and Flame, 159(2012) 3372–3397.
38. Cabra, R., Chen, J-Y., Dibble, R.W., Karpetis, A.N., and Barlow, R.S., "Lifted methane-air jet flames in a vitiated coflow", Combustion and Flame, 143(2005) 491–506.
39. O'Loughlin, W., and Masri, A.R., "The Structure of the Auto-Ignition Region of Turbulent Dilute Methanol Sprays Issuing in a Vitiated Co-flow", Flow Turbulence Combust. 89(2012) 13–35.
40. O'Loughlin, W., and Masri, A.R., "A comparative study of non-reacting and auto-igniting turbulent acetone sprays jet", in preparation.

41. O'Loughlin, W., "Investigations of auto-ignition in dilute spray flames", PhD Thesis, The University of Sydney, 2012.
42. Gordon, R.L., Masri, A.R., and Mastorakos, E., "Simultaneous Rayleigh Temperature, OH- and CH2O-LIF Imaging of Methane Jets in a Vitiated Coflow", Combust. Flame 155 (2008) 181–195.
43. Gordon, R.L., Masri, A.R., and Mastorakos, E., "Heat Release Rate as Represented by [OH]x[CH2O] and its Role in Autoignition", Combust. Theory and Modeling, 13 (2009) 645–670.
44. Kolmogorov, A.N. *Dokl. Akad. Nauk SSSR*. 66(1949) 825–828.
45. Sevik, M. and Park, S.H., "The splitting of drops and bubbles by turbulent fluid flow", Journal of Fluids Engineering. 95(1973) 53–60.
46. Kourmatzis, A. and Masri, A.R., "Multiple stage atomization of fuels for use in combustion applications", 18th Australasian Fluid Mechanics Conference, Tasmania, 3–7 December 2012.
47. Kourmatzis, A., Pham, P.X., and Masri, A.R., "Air assisted atomization and spray density characterization of ethanol and a range of biodiesels", Fuel, In Press, 2013.
48. Engelbert, C., Hardalupas, Y., and Whitelaw, J., "Breakup phenomena in coaxial airblast atomizers", Proceedings of the Royal Society A, 451(1995) 189–229.
49. Lasheras, J., Villermaux, E., and Hopfinger, E.J., "Break-up and atomization of a round water jet by a high speed annual air jet". Journal of Fluid Mechanics, 357(1998) 351–379.

A Comparative Study of the Simulation of Turbulent Ethanol Spray Flames

Colin R. Heye, Agisilaos Kourmatzis, Venkat Raman and Assaad R. Masri

Abstract Experimental data for a series of spray flames is utilized to perform analysis of validation studies conducted by multiple contributors. In this multiphase context, various choices for boundary conditions as well as modeling frameworks and formulations are evaluated. Both large eddy simulation (LES) and Reynolds-averaged Navier-Stokes (RANS) approaches showed the ability to capture droplet evolution with regards to mean and fluctuating velocities. This accuracy is contingent on the proper specification of both droplet and gas phase velocities at the jet exit. The combined effect of combustion and evaporation model choices impacts the downstream volume flux of droplets and resulting gas phase temperature. Further investigation is required to isolate individual model effects for high-temperature spray-laden environments. Proposed solutions involve the simulation of a wider array of flow conditions or lowerlevel experiments to remove the effects of model coupling.

1 Introduction

The simulation of turbulent spray combustion is inherently complex due to the multi-scale interactions between the turbulent flow, the combustion process, and spray droplet dispersion. While several models have been proposed, their validity has not been extensively tested using standard and canonical flame experiments. Establishing model validity is a difficult task, but is primarily attained by comparison with meaningful experimental configurations. In this sense, measurements made in flows that bear resemblance to the final problem of interest, say aircraft engines, is important. The focus of this work is such a validation exercise carried out using the so-called Sydney spray flame [23], which includes a set of experiments of a spray jet flame performed at a range of inflow conditions.

In this effort, we are guided by the success of the Turbulent Non-premixed Flames (TNF) workshop [9], where a similar validation exercise aimed at gas phase

C. R. Heye (✉) · V. Raman
Department of Aerospace Engineering and Engineering Mechanics,
The University of Texas at Austin, Austin, Texas, 78712, USA
e-mail: cheye@utexas.edu

A. Kourmatzis · A. R. Masri
School of Aerospace, Mechanical and Mechatronic Engineering,
The University of Sydney, Darlington, Australia

B. Merci, E. Gutheil (eds.), *Experiments and Numerical Simulations of Turbulent Combustion of Diluted Sprays,* ERCOFTAC Series 19, DOI 10.1007/978-3-319-04678-5_2,
© Springer International Publishing Switzerland 2014

turbulent combustion was developed. That program, in our opinion, led to a number of advances in the way research is conducted in the area of turbulent flames. First, it provided common ground for experimentalists and modelers to discuss the state-of-the-art in their respective areas, and to use this expertise to advance the goal of predictive modeling. The experimentalists became aware of the specific computational requirements, including well-characterized boundary conditions, which dramatically altered the quality of validation data. Similarly, modelers understood the intricate complexities in measurement science, which enabled better use of data and inference even with limited number of measurements. This workshop changed the modeling and validation perspective in a unique and positive way.

The Turbulent Combustion of Sprays (TCS) workshops have a similar mission but deal with the increased complexities arising from spray-based fuel injection. As we will discuss below, this presence of the liquid phase dramatically alters the level of information required for simulations, creating a tougher environment for developing well-characterized measurements. In particular, the inflow conditions require more detailed characterization than that envisioned in gas phase flames. The purpose of this article is twofold: (1) Provide a summary of the validation exercise conducted as part of the TCS2 and TCS3 workshops, and (2) discuss the key information needed to fully characterize and inform the simulations.

2 Experimental Configuration

In this section, the experimental configuration and a discussion of the validation data from the perspective of the simulations are provided.

The Sydney spray flame were conducted at University of Sydney by Masri and co-workers [5, 11, 23]. From a modeling perspective, this is one of the very few experiments that contain sufficient characterization of the configuration that allows for advanced modeling tools such as large eddy simulation (LES) to be used. The geometry of the burner is shown in Fig. 1. The burner consists of a round central jet, supplying atomized liquid fuel carried by air, a pilot flame and an outer air coflow. The burner is mounted vertically with a co-flow wind tunnel velocity of 4.5 m/s. The outer diameter of the burner annulus is 25 mm with a lip thickness of 0.2 mm. The pilot flame consists of a stoichiometric mixture of hydrogen, acetylene and air. The liquid spray is initially introduced 215 mm upstream of the nozzle exit plane. Previous spray flame experiments have used this burner, and additional details on the configuration can be found in [23, 37].

The series of experiments varied both liquid fuel and air mass flow rates in the core jet. Figure 2 shows the parameter space for the flame series, which lie well within the stability limits for ethanol fuels (shown by solid lines in Fig. 2). From within this extensive database, three flames were selected for further modelling studies, namely cases 2, 6 and 7 for either of acetone and ethanol fuels (i.e. Cases EtF2, EtF6, EtF7 and AcF2, AcF6, AcF7). In this initial work, only two flames are selected for further analysis, specifically the ethanol fuel cases EtF2 and EtF6. Flow parameters for these flames are seen in Table 1.

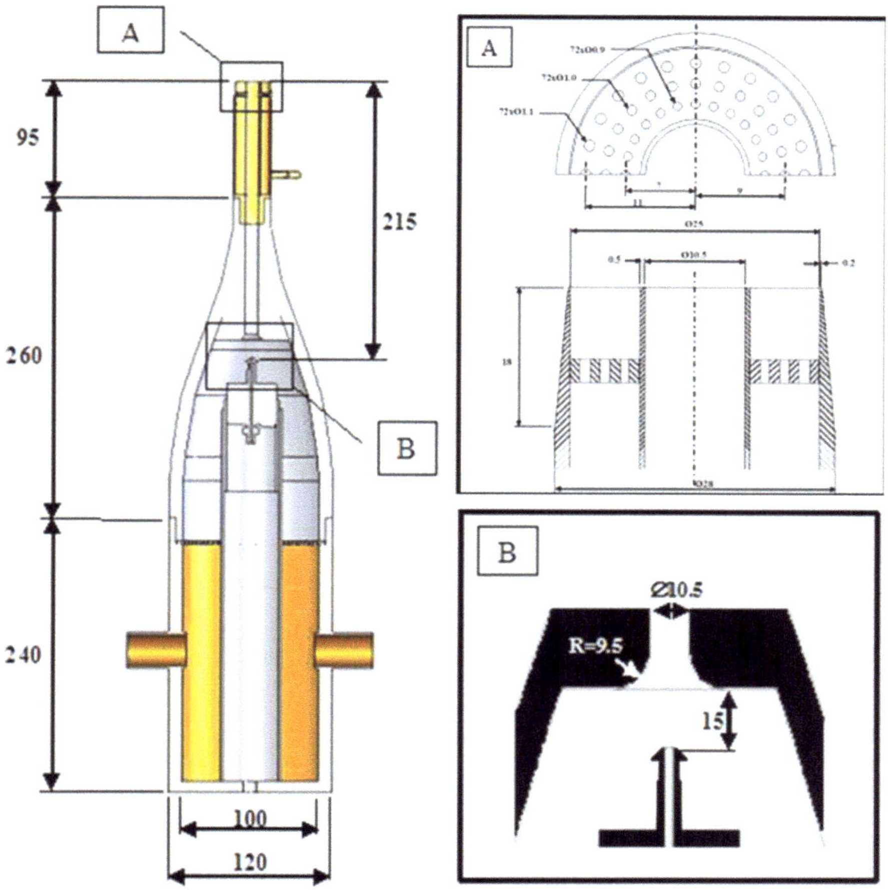

Fig. 1 Sydney burner configuration. Inset views show the pilot burner outflow (*top*) and nebulizer contour (*bottom*)

2.1 Inflow Conditions

As mentioned in Sect. 1, inflow conditions play a crucial role in the validation tests for spray flames. In aircraft engines, liquid jets are atomized before entering the main combustion region. In such flows, the description of the atomization process controls the spray number density function, as well as the spatial penetration of the fuel jet. Accurate characterization of the atomization process remains a challenging problem to model [10]. However, the focus of this work is on the interaction of atomized droplets with the turbulent flame, allowing for the use of simpler inflow conditions. In the experiments, a nebulizer is used to generate low-momentum droplets, which are then carried downstream by the surrounding flow. The droplet-air mixture passes through the 215 mm pipe, where both the gas phase turbulence and droplet velocities

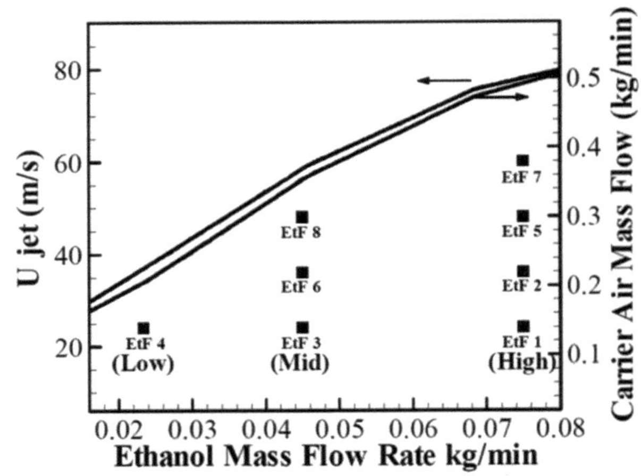

Fig. 2 Ethanol flame series air and fuel mass flow rate. Ethanol flame stability limits denoted by solid lines

Table 1 Selected flame conditions

	EtF2	EtF6
Bulk velocity (m/s)	36	36
Liquid fuel injection rate (g/min)	75	45
Measured liq. flow at jet exit (g/min)	66.3	41.1
Vapor fuel flow rate at jet exit (g/min)	8.7	3.9
Equivalence ratio at jet exit	0.3	0.2
Jet Reynolds number	30,500	27,400
Flame length (cm)	72	53

develop. Due to the relatively high volatility of ethanol, significant evaporation occurs in the pipe. This evaporation rate increases with fuel loading, and EtF2 exhibits much higher liquid-phase loss compared to EtF6 (Table 1).

The evaporation and the presence of the long pipe produce flow anisotropies, the impact of which are difficult to assess due to the lack of detailed measurements. First, due to significant droplet inertia, especially for the larger particles, it is reasonable to expect a change in the droplet size distribution as a function of the axial distance traversed in the pipe. Moreover, the presence of the walls could lead to additional issues including condensation and potential droplet-droplet interactions (although the bulk loading is still in the dilute regime). Second, evaporation inside the pipe leads to fuel release in the gas phase, which then mixes with the carrier fluid downstream of the nebulizer. Since the droplet evaporation itself is spatially inhomogeneous, gas phase fuel concentration is also expected to be inhomogeneous. Depending on the level of turbulence in the pipe, there could be incomplete mixing by the end of the pipe leading to a non-uniform spatially and temporally varying fuel inflow condition. At this point, the experiments do not have the information necessary to characterize the scalar distribution. Rather, any such discrepancy from the uniform inflow conditions has to be inferred from flame behavior downstream.

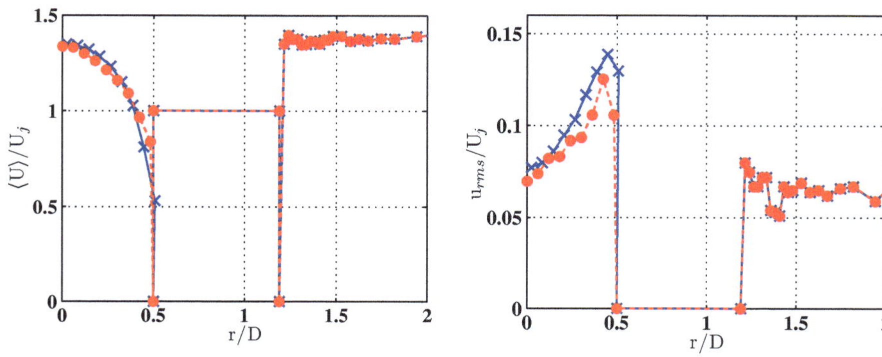

Fig. 3 Experimental gas phase streamwise mean (*left*) and RMS (*right*) velocities at $x/D = 0.3$ for cases EtF2 (—×—) and EtF6 (- -●- -)

The experimental data closest to the inflow consists of measurements taken at $0.3D$, where D denotes the diameter of the central jet. In order to validate simulation inflow conditions, experimental results at this most upstream location report radial profiles of two components of velocity for both the gas and dispersed phase, as well as droplet volume flux and probability density functions of droplet diameter. Figure 3 shows the streamwise mean and RMS velocities measured for the air/fuel mixture. Despite a significant difference in droplet mass loading between the two cases, the normalized mean gas velocities and relative turbulent intensities are similar. It should be noted that neither case is fully turbulent at the jet exit due to presence of droplets slowing the development of the near-wall turbulent structures.

From the simulation viewpoint, inflow conditions are required for both gas phase and droplet phase. While the gas phase properties are inferred based on the discussion above, specifying the inflow conditions for the droplet is more difficult. For the models used here, the droplet size distribution as well as the velocity distribution are required. The experimental data consists of droplet class-specific velocity component statistics at axial distance of $0.3D$ (Fig. 4). Clearly, there is appreciable variation in the velocity depending on the size, indicating that droplet residence time in the pipe will vary with diameter, thus causing droplets to be differentially evaporated. Smaller droplets will evaporate faster, leaving larger droplets behind as the spray traverses the pipe. This will result in a higher SMD further downstream with a distribution that is skewed towards larger droplets and an increase in D_{10} as is observed experimentally. Additionally, the slip velocity at the exit plane is not zero and larger droplets are slower than the gas phase, thus increasing their residence time. The impact of the skewed distribution and the slip velocities on the calculations is not known and needs to be ascertained.

Figure 5 shows the droplet size distribution function, normalized by the total number density. As expected the near-wall distribution contains more of the larger droplets, indicating that these droplets enter the boundary layer close to the pipe wall. This indicates that the number density function is nonuniform across the inlet.

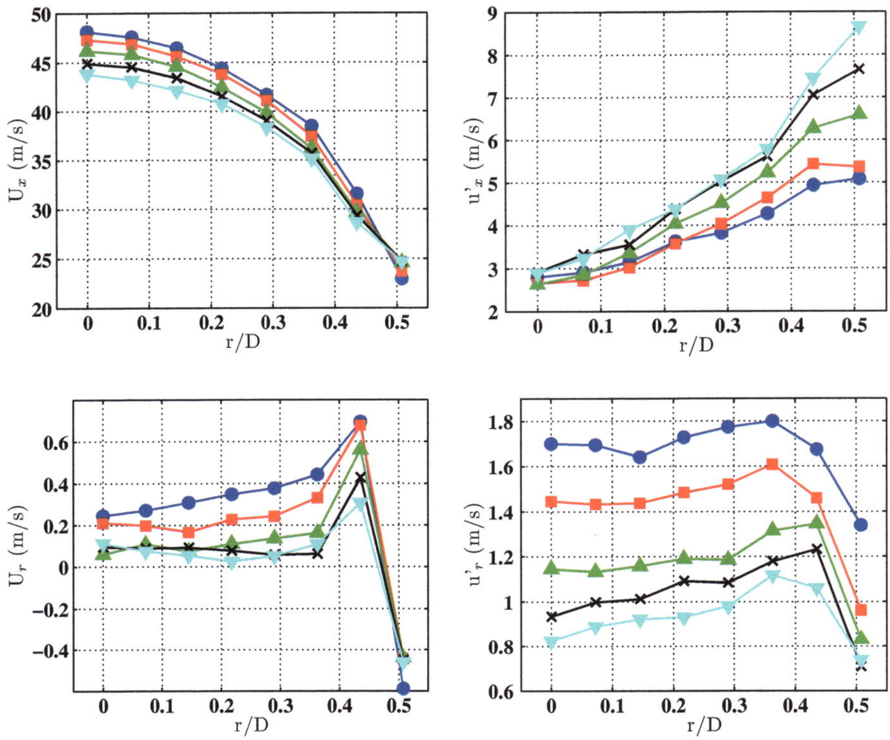

Fig. 4 Experimental droplet velocities from EtF2 at $x/D = 0.3$, including streamwise mean (*top left*), streamwise rms (*top right*), radial mean (*bottom left*), and radial rms (*bottom right*). Droplet diameter subranges are given by $0-10\ \mu m$ (–●–), $10-20\mu m$ (–■–), $20-30\ \mu m$ (–▲–), $30-40\mu m$ (–×–), and $40-50\mu m$ (–▼–)

The lack of a fully developed gas phase flow velocity (inferred from the smallest droplets) is likely due to the presence of these large droplets that attenuate the turbulent fluctuations.

2.2 Validation Data

In addition to the measurements at the nozzle exit, measurements were collected downstream in order to validate the unconstrained droplet mixing and evaporation processes. The available data includes radial profiles of the following time-averaged quantities: (1) Two component mean and RMS droplet velocities, (2) corresponding Reynolds stresses, (3) droplet volume flux, (4) D_{10} and D_{32} moments of the droplet diameter distribution, and (5) gas phase temperature. For these flames, gas phase temperature data is available at streamwise positions of $x/D = 10$, 20, and 30,

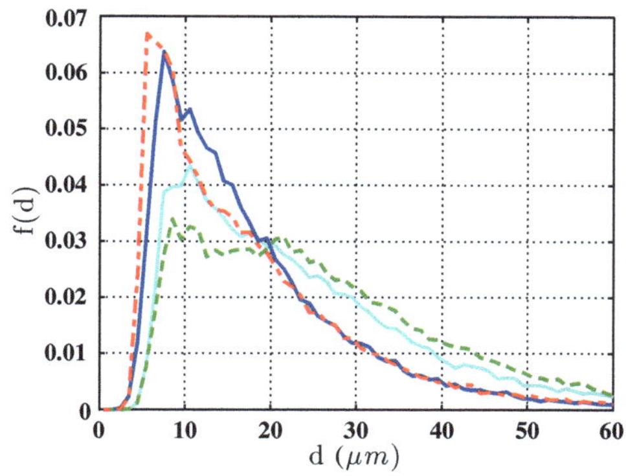

Fig. 5 Experimentally measured droplet diameter PDF at $x/D = 0.3$ at centerline for EtF2 (—) and EtF6 (- - -) and again near nozzle wall for EtF2 (—) and EtF6 (- -)

while all other quantities are available up to $x/D = 30$ in increments of five jet diameters.

3 Simulation Data from Multiple Contributing Groups

As part of the TCS3 workshop, four different groups contributed to the comparisons. The simulation approaches, inflow conditions, and combustion models run the gamut providing a rich set of simulation results for comparison. In this section, the individual contributions and the simulation techniques are discussed.

Of the four groups of contributors included in this study, three use LES and one uses a RANS approach. All four represent the dispersed phase with a coupled Lagrangian solver applying the point particle assumption [8, 38]. A summary of notable contributor methods are presented in Table 2 with the supplied data specified in Table 3, while additional simulations details and defining characteristics are provided in the following sections.

University of Darmstadt (UDRM) The contributions of M. Chrigui[1], S. Fernando, and A. Sadiki from Technische Unversitaet Darmstadt, designated here as UDRM, included all droplet related quantities of interest for case EtF6. Unique to this simulation amongst all the contributions, the domain extends upstream to the atomizer droplet injection point. The computational domain downstream of the nozzle exit extends 20.4 jet diameters radially and 76.2 jet diameters in the streamwise direction. The structured cylindrical mesh consists of 1.1 million cells. Additional details for analogous simulations have been previously published [6].

With the boundary conditions being specified at the upstream end of the pipe, both gas and droplet phase boundary conditions are relatively simple, allowing for the development of turbulent structures along the 215 mm pre-mixing length before the nozzle exit. The carrier air is introduced at a constant streamwise velocity across

[1] Currently holds a faculty position at Labouratoire des systemes embarqus et systemes nergtiques, ISSIG, Universit de Gabes

Table 2 Contributor configuration summary

	Chrigui et al.	Heye and Raman	Prasad et al.	Wunsch et al.
Designation	UDRM	UT-A	USYD	NMCA
Model framework	LES	LES	LES	RANS
Inflow location	Nebulizer exit	Nozzle exit	Nozzle exit	Nozzle exit
Grid size	1.1 million cylindrical cells	384 x 192 x 64 cylindrical	200 x 100 x 100 Cartesian	250 x 75 x 2
Gas phase inflow velocity	Bulk velocity	Fully turbulent pipe flow	Synthetic turbulent to match experimental mean and rms [21]	Prescribed experimental mean and rms
Droplet inflow velocity	Mean matches gas phase with Gaussian rms	Interpolate gas phase velocities to droplet locations	Match experimental mean and rms by diameter class	Match experimental mean and rms by diameter class
Droplet inflow diameter	Discrete MDF to match experimental PDF for droplet diameter classes	Log-normal distribution to match experimental centerline PDF	Nukiyama-Tanasawa distribution matching experimental centerline PDF [27]	Nukiyama-Tanasawa distribution respecting D_{10} and D_{32} for each diameter class [27]
Droplet inflow location	Gaussian random distribution	Uniform random distribution	Random with centerline bias to match experimental flux	Uniform random distribution
Evaporation model	Non-equilibrium [1, 36]	Equilibrium [34]	Equilibrium [1]	Non-equilibrium [25]
Combustion model	FPVA [3]	FPVA [32]	Detailed chemistry	Flamelet
Chemistry mechanism	56 species, 351 reversible step reaction mechanism [22]	50 species, 235 reaction combustion mechanism [4]	Reduced 27 step ethanol mechanism [30]	Ethanol oxidation mechanism [19]
Lagrangian SGS fluctuations	None	None	None	Turbulent dispersion model [24]

the radial profile, with no radial or swirl velocity and no initial turbulence. Droplets are initialized with a matching mean streamwise velocity, while including non-zero radial velocities in an attempt to replicate the nebulizer spray angle, as well as significant turbulent velocities. The droplet inflow diameters were tuned to obtain

Table 3 Summary of case-specific quantities of interest supplied by each contributor.

	Streamwise Vel.	Radial Vel.	D_{10}	Volume Flux	Gas Temp.
UDRM	EtF6	EtF6	EtF6	EtF6	
UT-A	EtF2/6	EtF2/6	EtF2/6		EtF2/6
USYD	EtF2	EtF2			EtF2
NMCA	EtF2/6	EtF2/6	EtF2/6	EtF2/6	EtF2/6

the proper nozzle outflow profiles with locations chosen from a Gaussian distribution to allow for centerline bias.

The continuous phase is solved following an Eulerian approach using a 3D low-Mach number LES code [20, 35, 40]. The solver uses block structured and boundary fitted grids. It is based on the finite volume method in which a co-located grid with a cell-centered variable arrangement is applied. For spatial discretization, specialized central differencing schemes that hold the second order for arbitrary grid cells are used. The convective term in the scalar transport is discretized using non-oscillatory, bounded total variation diminishing (TVD) scheme. For the time stepping, a multiple stage Runge-Kutta scheme with second order accuracy is used. A fractional step formulation is applied and at each stage a momentum correction is carried out in order to satisfy continuity. Hence, a Poisson equation is derived from the continuity equation and solved iteratively with multi-grid and SOR relaxation. The parcels are tracked using a Lagrangian approach, in which the ordinary differential equations describing movement in physical space, the temperature evolution and the evaporation-related mass loss are discretized using Euler first order schemes and solved explicitly.

Droplet evaporation is modeled using the uniform temperature (UT) model of Abramzon et al. [1] and Sirignano [36]. The model describes the evolution of the droplets' temperature and diameter, i.e. evaporation rate and energy flux through the liquid/gas interface. This model is based on the film thickness theory. It does not consider any temperature variation in the interior of the droplets (homogenous temperature). However, the temperature variation is time dependent and accompanied by an unsteady mass transition. Gas phase combustion was modeled using flamelet generated manifold approach [3] that utilizes a 56 species, 351 reaction ethanol combustion mechanism [22].

University of Texas at Austin (UT-A) The LES data collected by C. R. Heye and V. Raman (see [14] for details), further designated by UT-A, included all validation quantities except for droplet volume flux for both EtF2 and EtF6. The simulation inflow is located at the outflow of the jet nozzle, allowing for more direct application of the most upstream experimental measurements. The cylindrical grid is $192 \times 128 \times 64$ cells in the streamwise, radial and azimuthal directions, respectively. The domain captures the entire flame, extending 15 jet diameters in the radial direction and 80 diameters in the streamwise direction.

Gas phase inflow velocities were taken from a separate fully turbulent pipe simulation without spray to provide time-correlated turbulence. Droplet velocities at the inflow were determined by interpolating gas phase velocities to each droplet's location, thus imposing a small Stokes number approximation. The diameter of inflowing droplets was chosen from a log-normal PDF fit to experimental centerline measurements for each flame. This distribution was assumed to be valid at all radial locations within the core jet. Droplets were uniformly distributed across the jet.

The flows considered here fall in the low-Mach number regime but with variable density induced by boundary conditions and combustion-related energy release. Hence, a low-Mach number fractional time-stepping based LES solver is employed [2, 31, 32]. The essential components of the low-Mach number algorithm are the velocity advancement and pressure-based velocity correction to enforce continuity equation. In order to increase the time-step used, the viscous terms and the convection terms in the radial and azimuthal directions are treated implicitly [7]. To reduce computational expense, an iterative algorithm is used to solve the resulting nonlinear discretized equation [2, 31]. A second-order central scheme is used for spatial discretization in the momentum equations, while a third-order upwinded scheme [13] is used to discretize the nonlinear terms in the scalar transport equations. Further details of the LES algorithm are provided in [7]. The turbulent diffusivity and viscosity terms are modeled using a dynamic Smagorinsky approach [26].

An equilibrium evaporation model was used [34] without accounting for any sub grid turbulent dispersion. In order to improve accuracy in combustion modeling over the commonly used presumed PDF methods while maintaining computational efficiency, a transported joint scalar PDF method for spray laden flows has been developed and applied by the contributors [14]. The joint-PDF of mixture fraction and progress variable is solved using this approach, with the chemical source term for progress variable obtained from a lookup-table. The table itself was constructed based on the flamelet/prigress variable approach [32] using a 50 species, 235 reaction mechanism [4].

University of Sydney (USYD) The LES results submitted by V. N. Prasad, S. Navaro-Martinez[2] and K. H. Luo[3], designated henceforth as USYD, contain streamwise droplet velocities and gas phase temperature for EtF2. As with UT-A simulations, the domain inflow is specified at the pipe exit and includes a cross-section of 12 jet diameters in each spanwise direction, with a streamwise extent of 39 jet diameters. The Cartesian grid consists of $100 \times 100 \times 200$ points in the three coordinate directions.

The gas phase inflow velocities utilized the digital inflow generator of di Mare et al. [21] using the experimentally-measured mean and RMS velocities at the pipe exit. Droplet diameters are initialized at each time step through random selection from the reported centerline PDF form [27]. This PDF is presumed accurate across the entire jet cross-section. Each component of the droplet velocity is taken from the experimental measurements, with diameter subrange dependence enforced.

The in-house code BOFFIN [15] developed at Imperial College London was used for the LES simulation. The code is based on a semi-implicit low-Mach number formulation, with all spatial gradients being discretised with energy conserving second order schemes except for the scalar convection, for which a TVD scheme is applied. Following previous studies (e.g. [16, 17]), eight stochastic fields are employed to

---

[2] Imperial College London
[3] University of Southampton

characterise the turbulence-chemistry interactions at the sub-grid level and more details of the implementation of the stochastic fields can be obtained from [18].

The equilibrium droplet evaporation model [1] was used with no modeling of turbulent dispersion. The combustion modeling technique involved a direct calculation of the sub grid scalar PDF using a stochastic field approach including the effects of reaction source term fluctuations [39]. The chemical source terms were modeled using a reduced 27-step ethanol oxidation mechanism [30].

NUMECA Interational (NMCA) The RANS results provided by D. Wunsch, J. E. Anker, K. Claramunt and C. Hirsch at *NUMECA International* (Brussels, Belgium), designated as NMCA, included all validation quantities for both test cases except for temperature profiles for EtF6. In this RANS work, the domain is a 2D wedge of 5 degrees and a radian and axial extension of 15 and 50 jet diameters from the nozzle exit, respectively. The axisymmetric mesh consists of 250 $\times$ 75 $\times$ 2 nodes in the streamwise, radial and azimuthal directions, resulting in 18426 hexahedral cells. Grid convergence was ensured but is not shown here.

Gas phase boundary conditions are taken from the experimental data with fitted mean and rms profiles based on local turbulent intensity. The droplet velocity and Reynolds stress in axial direction initial conditions are taken from the experimentally measured data for size classes up to 50 microns. Larger size classes are released with the same velocity as the largest class measured. The droplet size distribution is initialized with Nukiyama-Tanasawa [27] distribution respecting D_{10} and D_{32} diameters for each of the distribution. Droplet diameters larger than 150 microns were neglected.

The simulations were carried out using FINE^{TM}/Open [28, 29] code, which is a based on a finite-volume discretization approach. A second order central scheme with artificial dissipation is used for spatial differentiation. Time integration is performed using a 4-stage Runge-Kutta scheme applying a multigrid technique and residual smoothing is used. Hakimi preconditioning [12] is applied to accelerate convergence for the present low-Mach number flow. The droplet trajectories are integrated with an exponential scheme from the analytical form of the droplet momentum equation.

Turbulence is modeled using the k-epsilon equations with a round jet correction [33]. Droplet trajectories are integrated with an exponential scheme and the non-equilibrium evaporation model of Miller et al. [25] is used in conjunction with a sub grid turbulent dispersion model [24]. Combustion is modeled using a RANS/flamelet approach with flame strain rates in the range $10 - 1000\ s^{-1}$. The mechanism considered for this study was from Konnov et al [19].

4 Results and Discussion

In the following sections, comparisons with experimental data are presented. While the overall objective of the validation study was to identify model shortcomings, the unique nature of spray systems introduces a number of difficulties that prevent a

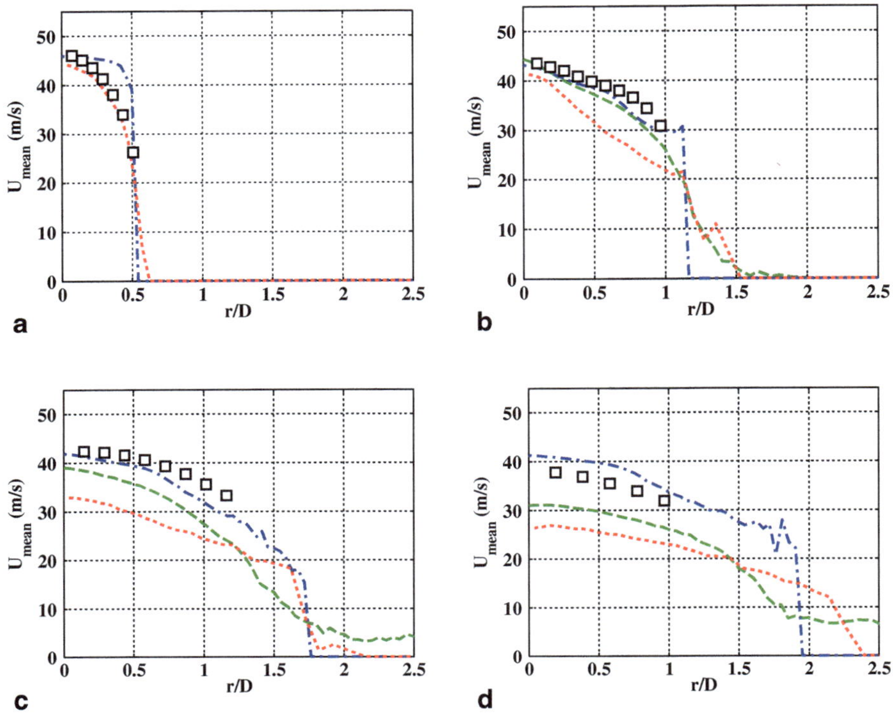

Fig. 6 Droplet streamwise mean velocities for EtF2 at $x/D = 0.3, 10, 20, 30$. Contributors given by UT-A (_._._), USYD (_ _ _), and NMCA (....), along with the experimental measurements (□)

direct evaluation of the combustion models. In this sense, this edition of the workshop has helped identify some common issues in simulating spray flames. Below, the discussion is centered around these validation issues. The two main validation quantities are the droplet characteristics, measured as droplet diameter statistics and mass flux, and gas phase temperature at select axial locations.

4.1 Spray Droplet Evolution

4.1.1 Droplet Velocites

All three contributors for case EtF2 prescribed the simulation inflow at the exit of the jet nozzle albeit with important differences. While USYD and NMCA determined the droplet inflow velocities directly from reported experimental measurements, UT-A simply interpolated gas phase velocities to the droplet locations. The impact of a fully turbulent inflow condition can be seen in Fig. 6, where even by $x/D = 0.3$, the results from UT-A show the streamwise droplet mean velocity retains the gas phase

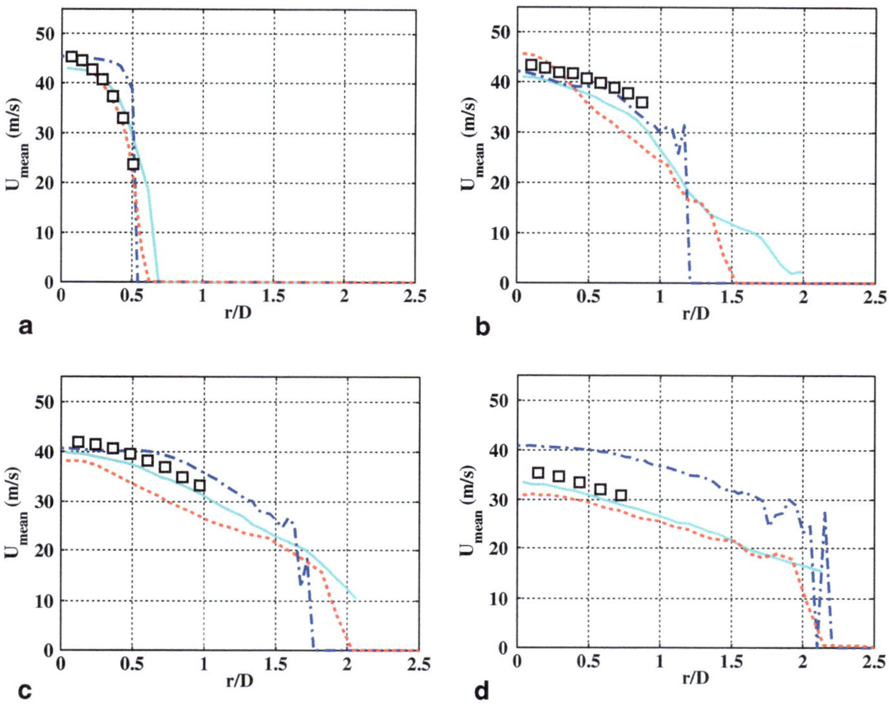

Fig. 7 Droplet streamwise mean velocities for EtF6 at $x/D = 0.3, 10, 20, 30$. Contributors given by UDRM (—), UT-A (-.-), and NMCA (....), along with the experimental measurements (□)

velocity profile. On the other hand, the RANS work by NMCA accurately reproduces the near field experimental streamwise mean. As the droplets progress downstream, however, the two simulations employing artificial turbulence boundary conditions predict much faster decay in droplet velocity. This over prediction of velocity decay may be an effect of relatively high droplet mass loading when compared to the results for EtF6 discussed below.

For case EtF6, the remaining liquid fuel at the nozzle exit is only 60 % of that for EtF2 as seen in Table 1. Although there is less evaporation in the upstream pipe, the lower droplet mass loading leads to decreased drag on the gas phase. In turn, this reduces the impact of droplet presence on turbulent evolution. The reduction in mass loading additionally reduces the overall equivalence ratio of the flow, which could lead to a weakened flame front characterized by intermittent local extinction. Consequently, re-laminarization through heat release is lessened in comparison to EtF2. Both these effects lead to a higher jet spreading rate combined with an enhanced decay of stream wise velocity, as seen in Fig. 7. The UDRM LES along with the RANS now predict the downstream profile much more accurately, while UT-A results maintain a relatively high mean velocity. These results are consistent with the experimental observations for these two flames. If all other factors are excluded,

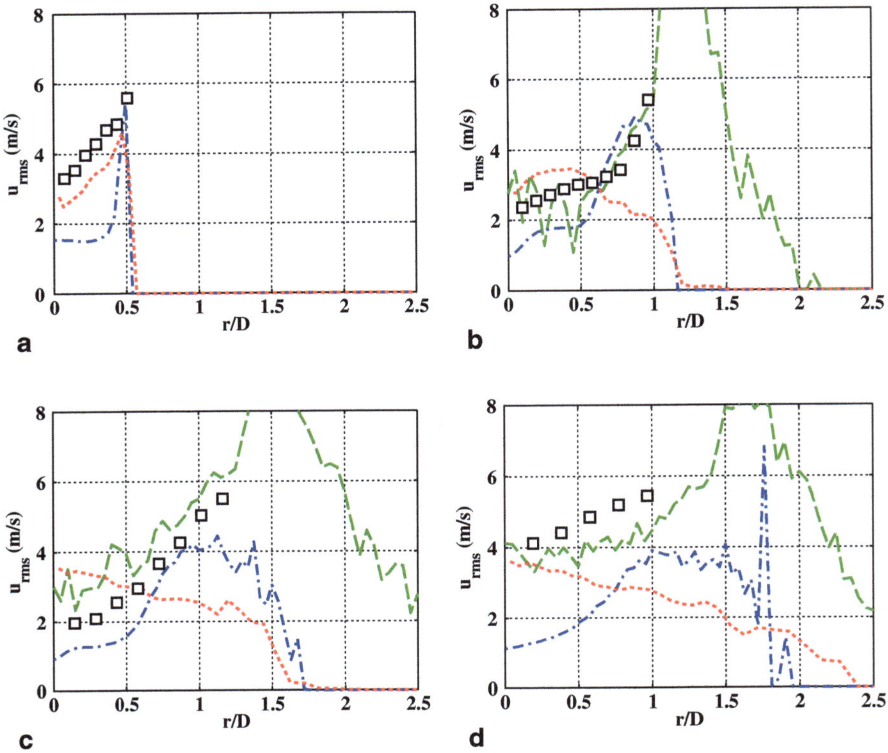

Fig. 8 Droplet streamwise RMS velocities for EtF2 at $x/D = 0.3, 10, 20, 30$. Contributors given by UT-A (-.-.-), USYD (– –), and NMCA (....), along with the experimental measurements (□)

it could be argued purely from the combustion modeling standpoint that the UT-A simulations over predict reactivity and the USYD/NMCA calculations under predict flame strength. Of course, due to the inherent coupling between numerous sub models such a statement cannot be made conclusively. Nevertheless, the sensitivity of the results to combustion model parameters is worth testing.

The streamwise RMS droplet velocities display some interesting results (Figs. 8 and 9). First, the LES computation that uses experimental conditions at $x/D = 0.3$ for inflow (USYD) generally produces better agreement with the experimental data. The UT-A computations that prescribe the spray inflow velocity based on the gas phase velocity field underpredicts the RMS quantity. Second, the RANS simulation indicates a reduction in RMS velocities at the outer edge of the droplet-laden central jet. This could be a result of the deficiencies of the RANS turbulence model in capturing the shear-flow turbulence or droplet-turbulence interactions. Consequently, it appears that LES predicts such interactions reasonably well if the inflow spray characteristics are taken from experiments. Interestingly, this trend is present for both flames.

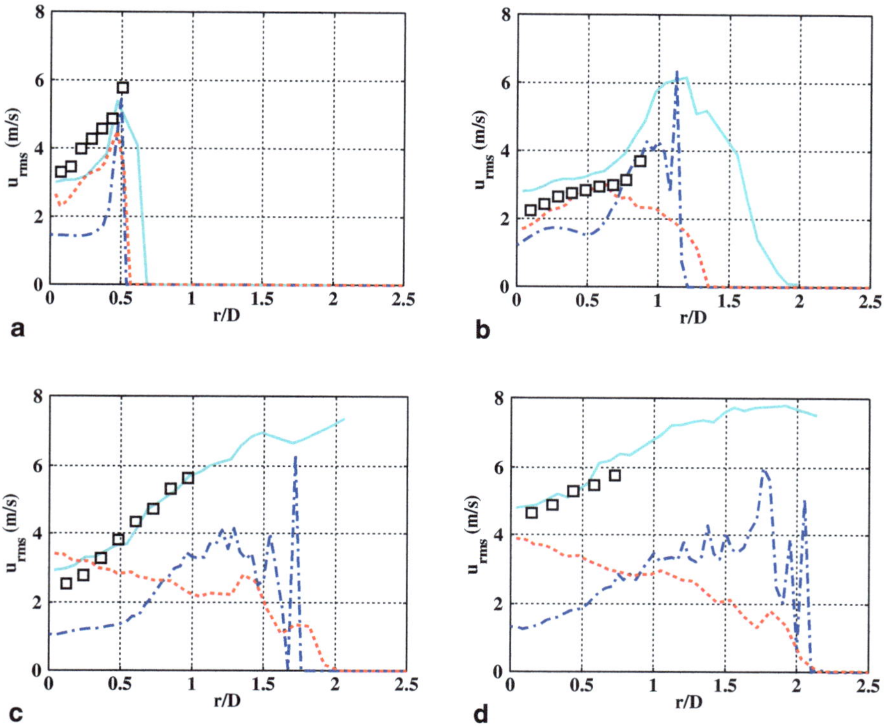

Fig. 9 Droplet streamwise RMS velocities for EtF6 at $x/D = 0.3, 10, 20, 30$. Contributors given by UDRM (—), UT-A (–·–), and NMCA (····), along with the experimental measurements (□)

Despite the significant impact of gas phase turbulence levels in the streamwise predictions of UT-A, the radial component of droplet velocity remains relatively independent and compares well to both the experimental results and the RANS work. Figures 10 and 11 show that fully turbulent LES captures both the near field and far field values very well. This result is also surprising given that the streamwise velocity statistics exhibit significant errors. It is very likely that this good agreement is merely coincidental given the large discrepancies in the streamwise component.

The evolution of the droplet radial RMS velocity (Fig. 11) shows contradictory results. The gas-phase based droplet properties used in the UT-A simulations seem to produce better agreement at the nozzle exit, while the upstream pipe based UDRM results show significant underprediction at this point. However, the UDRM results improve further downstream while the UT-A results seem to deteriorate with axial distance. Both simulations produce good agreement at the last downstream location. The RANS results from NMCA, on the other hand, consistently underpredict the RMS velocity, similar to the streamwise component results.

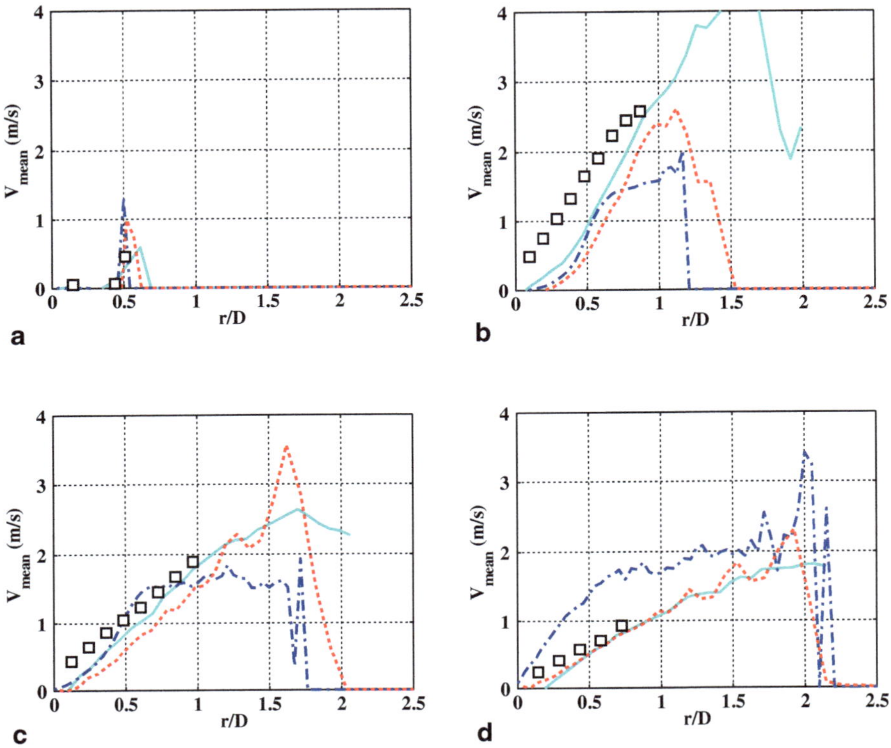

Fig. 10 Droplet radial mean velocities for EtF6 at $x/D = 0.3, 10, 20, 30$. Contributors given by UDRM (—), UT-A (-.-), and NMCA (....), along with the experimental measurements (□)

4.1.2 Evaporation Rates

Another useful metric for comparison is the droplet population size. Figures 12 and 13 show the evolution of the mean diameter. From the EtF2 results, it is seen that the UT-A simulations show considerable difference in trends, with high droplet diameter in the shear layer separating the central jet from the coflow. Both the NMCA and USYD simulations show much better qualitative agreement, and good quantitative agreement at intermediate distances from the nozzle exit. A similar behavior is observed in the UT-A simulation results for EtF6, while the UDRM LES and NMCA RANS results are in better agreement with the experimental data. The differences arise due to the differential evaporation based on droplet size distribution. The UT-A results indicate that the models used within allow smaller droplets to evaporate faster near the flame front, causing larger sized droplets to be left behind. While this is not inconsistent with basic theory, this behavior indicates that the evaporation model acting in conjunction with errors in the flame front description seems to lead to large discrepancies in the evolution of the droplet distribution. None

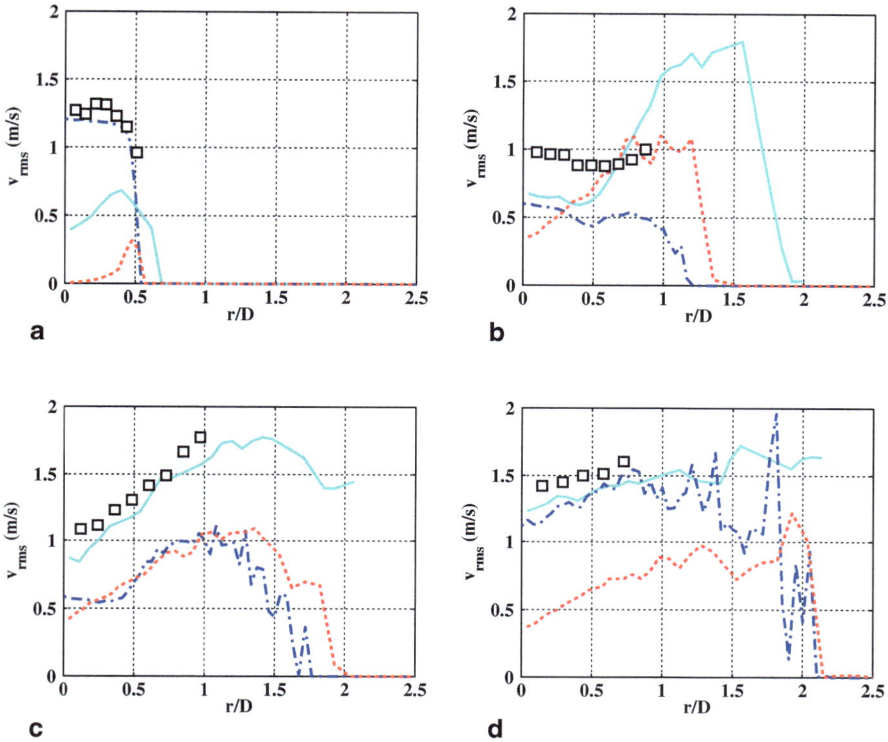

Fig. 11 Droplet radial RMS velocities for EtF6 at $x/D = 0.3, 10, 20, 30$. Contributors given by UDRM (—), UT-A (–·–), and NMCA (····), along with the experimental measurements (□)

of the models applied in this study, however, were able to capture the significant overall increase in mean diameter far downstream, especially near the jet centerline. For this to happen, smaller droplets in the core of the jet have to evaporate faster in a lower temperature carrier gas. Further analysis and research is needed to isolate the source of the incongruity.

The volume flux of droplets is another indirect measure of the evaporation rate. Only UDRM and NMCA provided data for this comparison (Fig. 14). Results shows that both LES and RANS match the boundary condition well, showing that both inflow definitions can accurately capture the global upstream evaporation. The LES results do show additional features such as increased flux near the centerline as well as a slight increase in flux at the edge of the shear layer. Further downstream, however, NMCA significantly overpredicts evaporation rates leading to reduced volume flux.

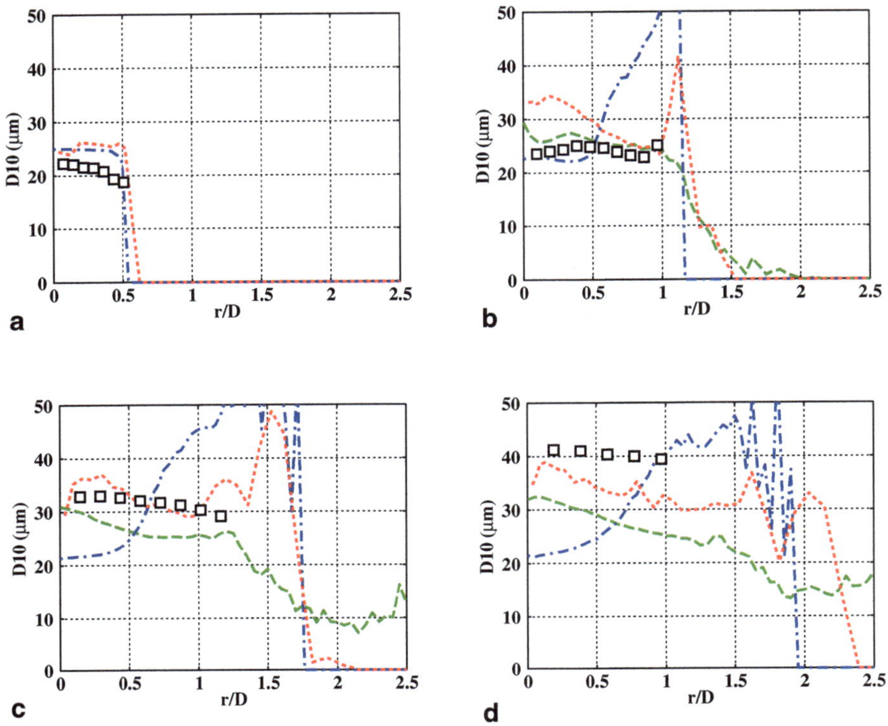

Fig. 12 Droplet mean diameter radial profile for EtF2 at $x/D = 0.3, 10, 20, 30$. Contributors given by UT-A (_._._), USYD (_ _ _), and NMCA (....), along with the experimental measurements (□)

4.2 Gas-phase Temperature

The validation data also provides gas phase radial temperature profiles at select axial locations. Figures 15 and 16 show the results for EtF2 and EtF6, respectively. It is seen that in the case of EtF2, the USYD calculations predict flame spread the most accurately amongst the three calculations, while UT-A and NMCA calculations under and over predict flame spread, respectively. In other words, the UT-A simulations show a flame that is not perturbed too much by the turbulent flow, leading to a locally confined flame front and significant extinction for EtF6. The NMCA simulations, as to be expected from RANS models, consistently predict a larger spread leading to higher temperatures further away from the centerline for both flame configurations. While the USYD calculations are more accurate, it is seen in general that the LES results underpredict temperature near the centerline. The higher temperature in the core of the flow could be the result of the flamelet model used in the NMCA calculations, which might over-predict reactions when even small amounts of fuel are present without an ignition source or enthalpy increase that is required by the FPVA

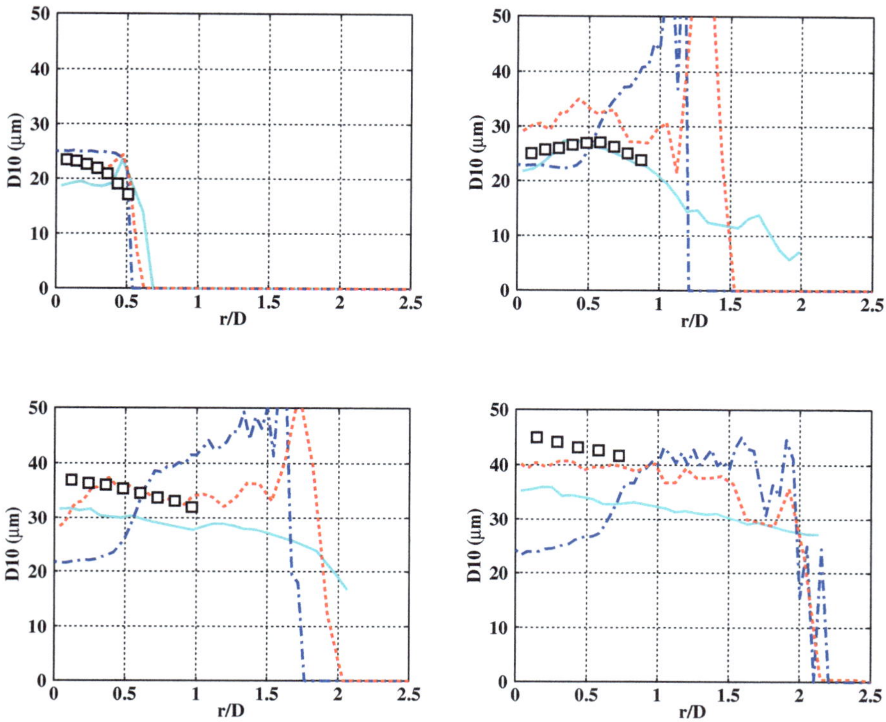

Fig. 13 Droplet mean diameter radial profile for EtF6 at $x/D = 0.3, 10, 20, 30$. Contributors given by UDRM (__), UT-A (_._._), and NMCA (....), along with the experimental measurements (□)

or the PDF method used in the LES calculations. In this sense, there is a direct link between the gas phase temperature and the combustion models used.

5 Concluding Remarks

In this study, a comparison of different simulations of the Sydney spray flame configuration was presented. The simulations spanned the entire spectrum of modeling approaches used. Consequently, although the data was rich in variability, it was also difficult to draw conclusive evidence from the comparisons. The following information was gleaned from these comparisons.

- Three different combustion models, namely the flamelet, FPVA, and PDF methods, were considered. It was found that the use of the flamelet model provided the best gas phase temperature predictions near the centerline, which is a direct consequence of the assumptions built into the flamelet description. The fact that the FPVA and PDF methods that require mixing of high-temperature fluid with the core of the jet to increase temperature leads us to conclude that the large scale

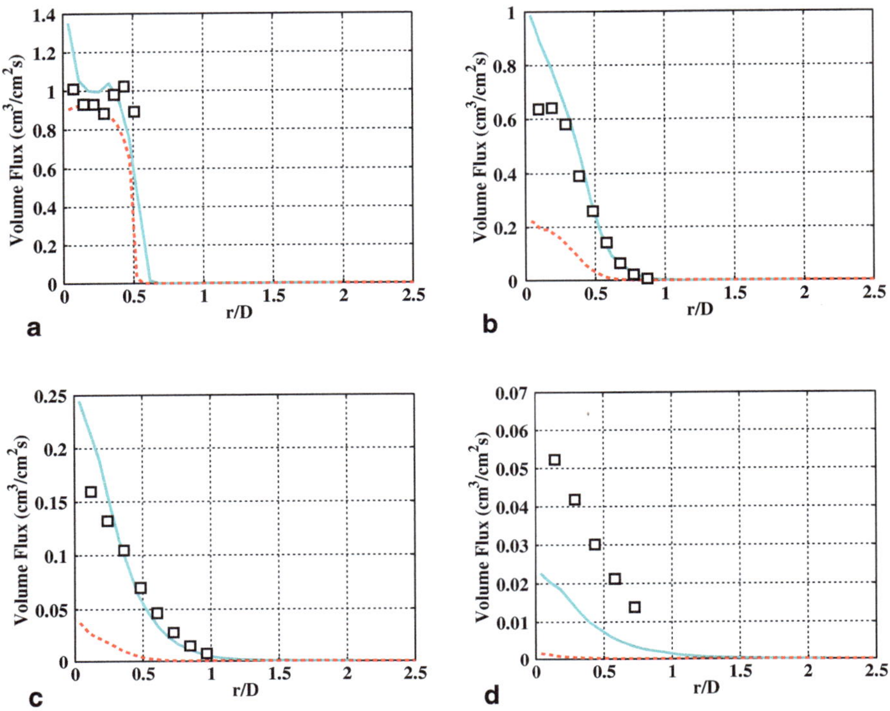

Fig. 14 Streamwise droplet volume flux for EtF6 at $x/D = 0.3, 10, 20, 30$ (top to bottom, left to right). Contributors given by UDRM (—) and NMCA (....), along with the experimental measurements (□)

mixing is still not correctly predicted by the simulations. This will in turn affect droplet evaporation and temperature evolution downstream. On the flip side, the higher temperature of the flamelet model causes very high droplet evaporation, as seen by the drastic underprediction of volume flux, albeit for a different flame condition.

- The ideal specification of simulation inflow conditions remains uncertain. While it seems advantageous to simulate the entire pipe rather than use the exit conditions for specifying the flame inflow conditions, the results do not indicate any major improvement. In fact, the UT-A calculations that use a simpler assumption by taking the droplet velocity properties directly from the gas phase seems to do as well as the other simulations in predicting the mean droplet properties. Of course, the velocity RMS seems to be more sensitive to the inflow conditions. Even in this regard, matching the experimental measurements at the location closest to the inflow seems to provide good results.
- The evaporation model is an important component of the modeling setup, but its effect cannot be discerned from this configuration directly. The main issue is the validity of the evaporation models in the vicinity of the flame, and it is not clear

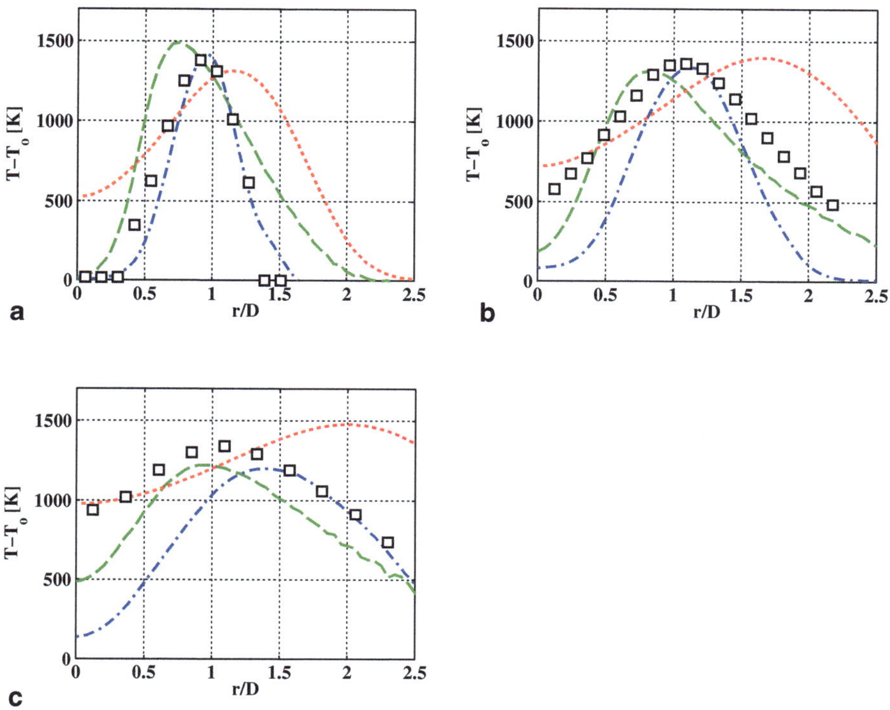

Fig. 15 Time-averaged gas phase temperature radial profile for EtF2 at $x/D = 10,20,30$. Contributors given by UT-A (_._._), USYD (___), and NMCA (....), along with the experimental measurements (□)

if such a precise question could be answered from the essentially very high level configuration. Nevertheless, this question needs to be explored computationally, mainly through analyzing sensitivity to existing models as a way of interpreting the impact of evaporation models on the results.

- There has been a general bias in the literature with regard to LES vs. RANS modeling approaches, favoring the former due to its ability to represent large scale mixing. We see some of these effects in this study, with the RANS approach failing to capture the RMS velocity and flame spread. However, if one would take into account the cost of the simulations, the LES approach does not seem to deliver vastly improved results that would warrant the unsteady three-dimensional computations. In this sense, LES modelers have their work cut out in demonstrating the validity of this approach for such complex flow problems.

This comparative study has also led to some intuition on the type of data that would be able to shed light on the modeling issues discussed above.

- The single biggest difference between gas phase and spray flames is the nature of the boundary conditions, due to the challenges associated with generating

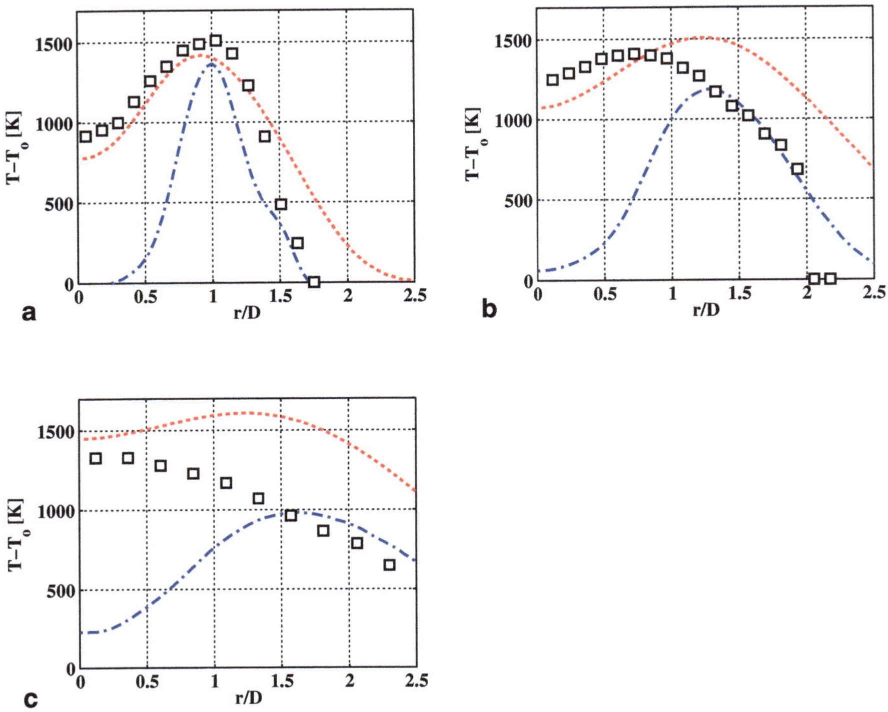

Fig. 16 Time-averaged gas phase temperature radial profile for EtF6 at $x/D = 10, 20, 30$. Contributors given by UT-A (_._._) and NMCA (....), along with the experimental measurements (□)

well-characterized spray populations. Toward improving inflow condition specifications, it would be supremely useful if detailed measurement of the fuel concentration profile as well as droplet evolution inside the development pipe is made available. In order to direct experimental investigations, a thorough numerical sensitivity study to the inflow parameters would also be beneficial. This would highlight parameters which should be the focus of future experiments for improving droplet inflow specifications. It is clear that in spite of the length of the pipe, the inertia of the droplets continues to play some role in their spatial distribution. In addition, fuel evaporation inside the pipe may not lead to homogeneous mixtures at the nozzle exit, which could severely alter the development of the flame front. Non-uniform droplet distribution will also impact the propagation of the flame in the near-entrance region.

- Given the flow complexity, it would be useful to identify and develop lower-level experiments that only couple a few of the physical phenomena in order to build confidence in the predictions. For instance, the aforementioned pipe information only couples droplet flow to evaporation without the interaction of the turbulent flow. Use of direct numerical simulation (DNS) or such sources of data will help isolate modeling problems.

- In comparing with experiments, it is important that the model is tested on multiple flames in the configuration rather than one. Due to the complexity of the problem, evaluating the sensitivity of the models to flow conditions rather than focusing on the quantitative prediction of a single flame condition is more illuminating.

Acknowledgments C. R. Heye and V. Raman were supported through an NSF CAREER Grant (NSF 0747427). A. R. Masri was supported by the Australian Research Council. The authors thank the contributors for providing the simulation data and for valuable discussion.

References

1. B. Abramzon, W. Sirignano, Droplet vaporization model for spray combustion calculations, International Journal of Heat and Mass Transfer 32 (9) (1989) 1605–1618.
2. K. Akselvoll, P. Moin, Large eddy simulation of turbulent confined coannular jets, Journal of Fluid Mechanics 315 (1996) 387–411.
3. R. J. M. Bastiaans, J. A. van Oijen, L. P. H. de Goey, Application of flamelet-generated manifolds and flamelet analysis of turbulent combustion, International Journal for Multiscale Computational Engineering 4 (3) (2006) 307–317.
4. Chemical-Kinetic Mechanisms for Combustion Applications, San Diego Mechanism web page, Mechanical and Aerospace Engineering (Combustion Research), University of California at San Diego (http://combustion.ucsd.edu). (2011).
5. Y. Chen, S. Starner, A. Masri, A detailed experimental investigation of well-defined, turbulent evaporating spray jets of acetone, International Journal of Multiphase Flow 32 (2006) 389–412.
6. M. Chrigui, A. Masri, A. Sadiki, J. Janicka, Large eddy simulation of a polydisperse ethanol spray flame, Flow, Turbulence and Combustion (2013) 1–20.
7. O. Desjardins, G. Blanquart, G. Balarac, H. Pitsch, High order conservative finite difference scheme for variable density low Mach number turbulent flows, Journal of Computational Physics 227 (15) (2008) 7125–7159.
8. J. K. Dukowicz, A particle-fluid numerical model for liquid sprays, Journal of Computational Physics 35 (1980) 229–253.
9. Eleventh International Workshop on Measurement and Computation of Turbulent Non-premixed Flames, Darmstadt, Germany, 2011. URL http://www.sandia.gov/TNF/11thWorkshop/TNF11.html
10. M. Gorokhovski, M. Herrmann, Modeling primary atomization, Annual Review of Fluid Mechanics 40 (2008) 343–366.
11. J. D. Gounder, A. Kourmatzis, A. R. Masri, Turbulent piloted dilute spray flames: Flow fields and droplet dynamics, Combustion and Flame 159 (11) (2012) 3372–3397.
12. N. Hakimi, Preconditioning methods for time dependent navier-stokes equations, Ph.D. thesis, Vrije Universiteit Brussels, Brussels, Belgium (1997).
13. M. Herrmann, G. Blanquart, V. Raman, A bounded quick scheme for preserving scalar bounds in large-eddy simulations, AIAA Journal 44 (12) (2006) 2879–2880.
14. C. R. Heye, V. Raman, A. R. Masri, LES/probability density function approach for the simulation of an ethanol spray flame, Proceedings Of The Combustion Institute 34 (Part 1) (2013) 1633–1641.
15. W. P. Jones, F. di Mare, A. J. Marquis, LES-BOFFIN: Users Guide, Tech. Rep., Imperial College London (2002).
16. W. P. Jones, S. Navarro-Martinez, Large eddy simulation of autoignition with a subgrid probability density function method, Combustion and Flame 150 (3) (2007) 170–187.
17. W. P. Jones, S. Lyra, S. Navarro-Martinez, Numerical investigation of swirling kerosene spray flames using Large Eddy Simulation, Combustion and Flame 159 (4) (2012) 1539–1561.

18. W. P. Jones, A. J. Marquis, V. N. Prasad, LES of a turbulent premixed swirl burner using the Eulerian stochastic field method, Combustion and Flame 159 (10) (2012) 3079–3095.
19. A. A. Konnov, R. J. Meuwissen, L. P. H. de Goey, The temperature dependence of the laminar burning velocity of ethanol flames, Proceedings of the Combustion Institute 33 (Part 1) (2011) 1011–1019.
20. T. Lehnhauser, M. Schafer, Improved linear interpolation practice for finite-volume schemes on complex grids, International Journal for Numerical Methods in Fluids 38 (7) (2002) 625–645.
21. L. di Mare, M. Klein, W. Jones, J. Janicka, Synthetic turbulence inflow conditions for large-eddy simulation, Physics of Fluids 18 (2).
22. N. Marinov, A detailed chemical kinetic model for high temperature ethanol oxidation, International Journal of Chemical Kinetics 31 (3) (1999) 183–220.
23. A. R. Masri, J. D. Gounder, Turbulent Spray Flames of Acetone and Ethanol Approaching Extinction, Combustion Science and Technology 182 (4–6) (2010) 702–715.
24. J.-P. Minier, E. Peirano, The PDF approach to turbulent polydispersed two-phase flows, Physics Reports 352 (2001) 1–214.
25. R. Miller, K. Harstad, J. Bellan, Evaluation of equilibrium and non-equilibrium evaporation models for many-droplet gas-liquid flow simulations, International Journal of Multiphase Flow 24 (6) (1998) 1025–1055.
26. P. Moin, K. Squires, W. Cabot, S. Lee, A dynamic subgrid-scale model for compressible turbulence and scalar transport, Physics of Fluids A 3 (1991) 2746–2757.
27. S. Nukiyama, Y. Tanasawa, Experiments on the atomization of liquids in an air stream, Transactions of the SME Japan 6 (1940) 5–7.
28. NUMECA International, Belgium, Theoretical Manual FINE/Open 212.3 (2013).
29. NUMECA International, Belgium, User Manual FINE/Open 212.3 (2013).
30. M. Okuyama, S. Hirano, Y. Ogami, H. Nakamura, Y. Ju, H. Kobayashi, Development of an Ethanol Reduced Kinetic Mechanism Based on the Quasi-Steady State Assumption and Feasibility Evaluation for Multi-Dimensional Flame Analysis, Journal of Thermal Science and Technology 5 (2) (2010) 189–199.
31. C. D. Pierce, Progress-variable approach for large-eddy simulation of turbulence combustion, Ph.D. thesis, Stanford University (2001).
32. C. D. Pierce, P. Moin, Progress-variable approach for large-eddy simulation of non-premixed turbulent combustion, Journal of Fluid Mechanics 504 (2004) 73–97.
33. S. B. Pope, Explanation of Turbulent Round-Jet-Plane-Jet Anomaly, AIAA Journal 16 (3) (1978) 279–281.
34. J. Reveillon, L. Vervisch, Analysis of weakly turbulent dilute-spray flames and spray combustion regimes, Journal of Fluid Mechanics 537 (2005) 317–347.
35. A. Sadiki, M. Chrigui, J. Janicka, M. Maneshkarimi, Modeling and simulation of effects of turbulence on vaporization, mixing and combustion of liquid-fuel sprays, Flow Turbulence and Combustion 75 (1-4) (2005) 105-130, 2nd International Workshop on Trends in Numerical and Physical Modelling of Turbulent Processes in Gas Turbine Combustors and Automotive Engines, Univ Heidelberg, Heidelberg, GERMANY, APR 01-02, 2004.
36. W. A. Sirignano, Fluid dynamics of sprays - 1992 freeman scholar lecture, Journal of Fluids Engineering 115 (1993) 345–378.
37. S. Starner, J. Gounder, A. Masri, Effects of turbulence and carrier fluid on simple, turbulent spray jet flames, Combustion and Flame 143 (4) (2005) 420–432.
38. S. Subramaniam, Lagrangian-Eulerian methods for multiphase flows, Progress in Energy and Combustion Science 39 (2-3) (2013) 215–245.
39. L. Valino, A field Monte Carlo formulation for calculating the probability density function of a single scalar in a turbulent flow, Flow Turbulence and Combustion 60 (2) (1998) 157–172.
40. B. Wegner, A. Maltsev, C. Schneider, A. Sadiki, A. Dreizler, J. Janicka, Assessment of unsteady RANS in predicting swirl flow instability based on LES and experiments, International Journal of Heat and Fluid Flow 25 (3, SI) (2004) 528–536, 3rd International Symposium on Turbulence and Shear Flow Phenomena (TSFP-3), Sendai, JAPAN, JUN 25–27, 2003.

Evaporation Modeling for Polydisperse Spray in Turbulent Flow

Mouldi Chrigui, Fernando Sacomano, Amsini Sadiki and Assaad R. Masri

Abstract Based on an overview of existing vaporization models, a suggestion for capturing phase transition in a turbulent two phase flow is made. Focus is put on the Uniform Temperature Model (UTM). Comparison between equilibrium and non-equilibrium evaporation models to experimental data is highlighted. Two configurations with different fuels, i.e. different thermodynamic properties, are investigated and the results of both models are validated with the measurements. The configurations exhibit completely different boundary conditions and polydisperse turbulent multiphase flows with different classes and probability distribution of the droplet diameters. Large eddy simulation (LES) and Reynolds averaged numerical simulation (here RANS) models are used to capture the turbulence. In both configurations, results show that non-equilibrium effects influence the vaporization significantly. The UTM with the extension of non-equilibrium, by Langmuir and Knudsen, capture the vaporization well, whereas the equilibrium model over-predicts the volume flux of the liquid phase, i.e. the vaporization process is developing slower in case of equilibrium model. Worth to notice that the mean droplet diameter is between 20 and 40 μm. Thus the ratio of surface to volume is important if compared to larger droplets. Non-equilibrium effects are then correspondingly important and the equilibrium model is not able to describe the phase transition process well.

M. Chrigui (✉) · F. Sacomano · A. Sadiki
Department of Mechanical Engineering, Institute for Energy and Powerplant Technology, Technische Universität Darmstadt, Petersenstr. 30, 64287 Darmstadt, Germany
e-mail: mchrigui@ekt.tu-darmstadt.de

M. Chrigui
Research Unit Materials, Energy and Renewable Energies, University of Gafsa, Gafsa, Tunisia

A. R. Masri
School of Aerospace, Mechanical and Mechatronic Engineering, The University of Sydney, 2006 Sydney, NSW, Australia

B. Merci, E. Gutheil (eds.), *Experiments and Numerical Simulations of Turbulent Combustion of Diluted Sprays,* ERCOFTAC Series 19, DOI 10.1007/978-3-319-04678-5_3,
© Springer International Publishing Switzerland 2014

1 Introduction

In combustion systems fueled by liquid fuel, the evaporation represents a decisive process of the fuel preparation. If evaporation is captured inaccurately, then the prediction of the mixing and the combustion processes would be necessarily not correct. The error transmission from the phase transition to the combustion variables is essentially influenced by the prediction of the droplet evaporation rate. Unfortunately a lot of parameters do affect the modeling of the evaporation, e.g. gas phase properties, spray injection, disperse phase characteristics, turbulence, modeling of fuel thermodynamic properties, etc. Numerous theoretical studies in literature have been carried out for the description of droplet evaporation.

From these studies, various droplet vaporization models have been suggested [1, 3, 7, 8, 11–15, 17, 19, 20, 23, 28, 30, 32]. They require physical description of the gas phase around the droplet and the transport processes inside the droplet (liquid phase) as well as the interfacial interactions of the two-phases. Depending on how the transport inside the droplet is described, the liquid phase models can be classified in five groups:

a. The infinite diffusion model,
b. The pure diffusion model,
c. The vortex model,
d. The simplified model, and,
e. The thin skin model.

The infinite diffusion model assumes that the transport processes of mass and heat occur so quickly that the physical properties and the chemical composition are always uniform inside the droplet. The pure diffusion model assumes that the transport is dominated by the molecular diffusion and that the property distribution is spherically symmetrical. The vortex models describe the droplet as unsteady and axisymmetric. The vortex motion inside the droplet is described by additional vorticity and stream function equations or analytical solutions including therefore the effects of both diffusion and convection. The simplified vortex model (equivalent to the effective diffusion models as compromise between (a) and (b)) attempts to account for the primary phenomena of droplet vaporization with less computational cost (compared with the vortex models). The thin skin model uses an infinitely thin layer of the liquid around the droplet [11]. This thin layer, which is mimicked by a thin skin, is heated and vaporized whereas the core of the droplet remains identical to the injection conditions. This model is suited for fuels, which demonstrate weak heat conductivity factors and/or a droplet temperature close to the boiling point. The heat transport is likely to occur in a steady manner within the droplet and transient effects are neglected. The thin skin model, generally, produces faster vaporization rate than other models.

Faeth [11] considered in his work a simple approach to compute the evaporation and combustion of sprays. He applied the d^2 model which has been widely used to describe the evaporation of fuel droplets. In this model droplets are heated till boiling temperature without any mass transfer. Then an evaporation period at an almost constant temperature takes place. This model is suitable in case of small heating period time compared to the droplet life time. The second type of evaporation models is the Uniform Temperature model (UT model) which neglects the mass transport inside the droplet. Here the temperature variation in the interior of the droplet is homogenous. It has an unsteady behavior and is accompanied with mass transition [15]. Both models do not consider any temperature gradient at the interior of the droplet. Therefore they do not need more computing time. In many cases, only the standard $k - \varepsilon$ model has been coupled to evaporation models, in particular to equilibrium evaporation models [14]. The implication of non equilibrium models has been accomplished in [19].

All models mentioned above are developed into low pressure conditions, i.e. $p_{atm} \leq p < 10$ bars. Oefelein et al. [20] tried to show the differences between classical low-pressure and high-pressure evaporation models. They mentioned that the subsequent (high pressure) drop regression process is different from that in the subcritical (low-pressure) state. Prommersberger et al. [23] built an experimental setup where evaporation of free falling monodisperse droplets was investigated at high pressure; they then compared the experimental results with numerical calculations based on some equilibrium droplet evaporation models. The convective transport of heat and mass at the droplet surface was calculated according to the film theory of Sirignano [28] accounting for the molar mass fraction through the Clausius-Clapeyron equation. He determined the best correspondence with experiments fit with the conduction limit model of Sirignano [28], which assumes a diffusive heat and mass transport within the droplet. Due to its high computational time consumption, this model is not recommended for large numbers of droplets.

Miller et al. performed in [19] an evaluation of existing evaporation models which are applicable to describe droplets having various diameters at a low pressure. Of particular interest was the vaporization of small single-component water, benzene, decane, heptane and hexane droplets in high temperature environments as found in many spray mixing and spray combustion processes. They noted that non-equilibrium effects became significant when the initial droplet diameter is less than 50 μm and that these effects are enhanced with increasing slip velocity. Thereby, they applied the non-equilibrium model by Langmuir-Knudsen [1]. The results agreed most favorably with a wide variety of experimental results.

In case of multi-component fuels, the UTM is generally extended to the Rapid Mixing Model RMM which implies an infinitely fast mixing of the fuel component, i.e. infinite mass diffusivity. This assumption implies that no gradients in temperature or concentration are considered within the droplet. The concept of multi-component evaporation is based on the fact that the liquid phase Lewis number, representing the ratio of thermal to mass diffusivity, could vary considerably with respect to the different component. Therefore, when Lewis number is high, the droplet interior will be heated at a rate much faster than that with which the more volatile components in the inner core can be transported to the surface where they are preferentially vaporized [17, 32]. The rapid mixing model assumes equilibrium evaporation of the droplet comparable with a batch distillation process [22]. The model delivers reasonable results for slow evaporation processes, when droplet internal heat conduction and diffusion does not have a major effect on the internal temperature and concentration profiles. The second type of multicomponent droplets is the diffusion limit model where the temperature and concentration distribution inside the droplet are determined using a discretization scheme to capture the heat conduction and mass diffusion. The disadvantage of this model is the enormous computational time [22].

The evaporation of complex fuels in industrial configuration have been so far very rarely simulated using LES due to the computational overhead originated by the large number of species. The acquisition of the different species and their distribution is indeed very important for the combustion since they exhibit different impacts depending on the application. Belan et al. [2], applied LES for the simulation of evaporating non-reactive multicomponent (MC) two phase flow. The species that constitute the real fuel, e.g. Diesel or Jet A are accounted for using a Probability Density Function (PDF) in the composition space. The chosen fuel composition however, was fitted as a single-gamma PDF, which indicates a limited range of species. Multicomponent evaporation has been also modeled in [13] and [25] using statistical methods that differentiate between species depending on their molar mass. Maqua et al. [17] performed an experimental and numerical investigation of bicomponent droplet evaporation. The emphasis was placed on the evaporation of binary mixture in hot air plume where a discrete component model was used. The application however, was far from the industrial standards. Sirignano and Wu [29] presented a generalized theory for multicomponent liquid fuel evaporation in arbitrary geometry. Though they made the assumption of non-forced or natural convection, they have not stated the way of capturing the turbulence of the carrier phase. Zugasti and Rosner [32] applied a MC droplet evaporation model to simulate JP4 kerosene. For the continuous mixture the spectral method was chosen. The results of mole fraction PDF have shown significant discrepancies. Bini and Jones [6] published a work that investigated the evaporation of one component fuel, i.e. acetone spray, which issues into an air co-flow to form an evaporating droplet jet using LES simulation. They adopted a probalistic description of the liquid phase and stochastic models for the filtered Lagrangian evaporation rates and heat release.

Senoner et al. [26] also used LES for the numerical investigation of evaporating two-phase flow in the so called MERCATO combustor which features a complex geometry. To mimic the fuel properties of the MC fuel used, a one component surrogate has been chosen. Lederlin and Pitsch [16] published a comparative work in which they also simulated the MERCATO configuration as previously investigated in [26] and [24] using LES and evaporating MC fuel. They restricted themselves on two components surrogate for jet A.

Pera et al. [21] modeled the SGS for the mixture of evaporating one component droplet. They used Euler-Euler (E-E) method. The modeling of the spray evaporation was not easy to capture even for single component. Burger et al. [7] developed a distillation curve model that is described by a single process variable which is based on the molar weight. They performed a RANS based simulation for Jet-A1 fuel. Unfortunately no information concerning the turbulence capturing of the gaseous phase was provided. Although the achieved results for the d^2-ratio showed good agreements compared with experimental data, the other droplet variables e.g. vapor mass flow, temperature were compared to the diffusion limit model.

Beside the above mentioned effort, valuable other research endeavour is put on the modeling and investigation of evaporating polydisperse turbulent spray under industrial operating conditions. In this study, focus is on the partial pre-vaporization of two volatile fuels using equilibrium and non-equilibrium models.

In the following, the concept of the uniform temperature evaporation model is presented (Sect. 2). The different assumption, needed to capture phase transition is listed and the non-equilibrium effects are introduced. In Sect. 3 the experimental configurations under investigation are presented followed by the description of the numerical framework used for this purpose (Sect. 4). The results are then provided and discussed (Sect. 5). The last section is devoted to conclusions and closing remarks.

2 Concept of the Uniform Temperature Model extended to Non-Equilibrium Effects

In this section two evaporation models, that are not complicated to implement and do not demonstrate a huge computational time, will be introduced. The first one is the so-called uniform temperature model by Sirignano [28]. This model represents an equilibrium evaporation model based on the film thickness theory. The second one is a non-equilibrium evaporation model developed by Langmuir and Knudsen (see [19] and [1]).

In order to ensure a mathematical description of the evaporation process, one should reduce the complexity of the theoretical description. Thus the following basic assumptions and simplifications are made [8]:

- One component model is considered, so that one solely deals with the so-called infinite conductivity model.
- Droplets are assumed to be spherical.
- Secondary atomization and coalescence of droplets are neglected as focus is on the dilute spray region. In other words simple elastic collisions between droplets and wall are assumed without any kind of film formation.
- Influence of the surface tension is neglected and a uniform pressure around the droplet is assumed.
- Uniform physical properties of the surrounding fluid and liquid-vapor thermal equilibrium on the droplet surface are assumed.
- The ambient air is not soluble in the droplet fluid.
- Chemical reactions and radiation are not considered.

In order to simplify the modeling of the evaporation process, the evaporation process is subdivided into three spatial zones (Fig. 1):

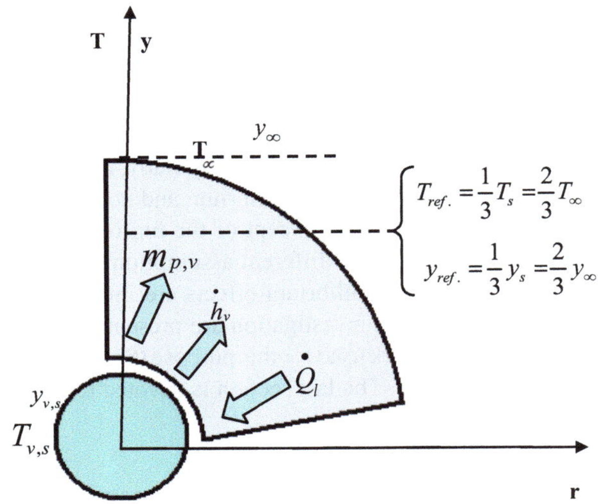

Fig. 1 Schematic description of evaporation

- The ambient gas phase, which represents the region infinitely far from the droplet. The temperature and vapor concentration are those of the carrier phase.
- The droplet interior.
- The liquid/gas interface, which represents the governing transport processes. In this region one observes radial mass diffusion, heat convection and conduction as well as forced convection.

The transport processes linked to the forced convection from the gas flow around the droplet, (located at the liquid/gas interface as presented above), is the fundamental difficulty in the development of practical evaporation models. For solving this problem, one has to start from an isolated droplet in a stagnant gas atmosphere with spherical symmetric transport of mass and energy. The effect of forced convection is then taken into account by means of empirical correlation factors (modified Nusselt and Sherwood numbers), as they will be introduced later on.

2.1 *The Uniform Temperature Model*

The Uniform Temperature model (equilibrium) does not consider any temperature variation in the interior of the droplet (homogenous temperature). However, the temperature variation has an unsteady behavior and is accompanied with mass transition [15]. As this model does not account for the gradient at the interior of the droplet, droplets are not discretized. Therefore this model does not require high computing time. The UT model describes the evolution of the droplet temperature and diameter, i.e. evaporation rate and energy flux through the liquid/gas interface (Fig. 1).

The vaporization rate is calculated by considering the mass transfer around the droplet following [8] and [4]:

$$\dot{m}_{p,v} = 2\pi r_p \overline{\rho_m} \overline{D_m} Sh^* \frac{\ln(1 + B_M)}{B_M} \tag{1}$$

where r_p is the droplet radius, $\overline{\rho_m}$ and $\overline{D_m}$ are the averaged values of the mixture density and binary diffusion coefficient throughout the film, respectively. B_M represents the Spalding's mass transfer number defined by

$$B_M = \frac{y_s - y_\infty}{1 - y_s}, \tag{2}$$

in which y_s is the surface vapor mass fraction and y_∞ is the vapor mass fraction far from the droplet. In particular y_s depends on the vapor relative pressure which itself depends on the droplet surface temperature. It is given as:

$$y_s = \frac{\upsilon_{s,eq}}{\upsilon_{s,eq} - (1 - \upsilon_{s,eq})\theta_2} \tag{3}$$

where θ_2 is the ratio of molecular weights. The molar mass fraction $\upsilon_{s,eq}$ is related to the saturation pressure through the Clausius-Clapeyron equation. Therefore it is referred to as an equilibrium model.

The accuracy of the evaporation rate depends strongly on the determination of the values of $\overline{\rho_m}$ and $\overline{D_m}$. Physical properties of the air vapor mixture (in the gaseous film around the droplet) are determined using the reference temperature and mass fraction. They are calculated using the Simpson or Sparrow and Gregg "1/3 rule" [4], from which the best accuracy of these quantities have been obtained (see Fig. 1) in comparison to the "2/3" rule. The droplet radius is obtained from the equation of the diameter evolution for each droplet by:

$$\frac{dD_p}{dt} = -\frac{2\dot{m}_{p,v}}{\pi \rho_L D_p^2} - \frac{D_p}{3\rho_L} \frac{\partial \rho_L}{\partial T_p} \frac{dT_p}{dt}, \tag{4}$$

where ρ_L is the liquid density, D_p the droplet diameter and T_p the droplet temperature.

Effects of convection on the vaporization and the heat flux rate are taken into account by means of semi-empirical correlations such as those for the drag coefficient, the Sherwood number and the Nusselt number. The quantity Sh^* denotes the modified Sherwood number which includes the effects of the Stefan flow. It is defined by

$$Sh^* = 2 + \frac{Sh_0 - 2}{F_M} \tag{5}$$

where

$$Sh_0 = 1 + \left(1 + \mathrm{Re}_p Sc_p\right)^{1/3} f(\mathrm{Re}_p) \tag{6}$$

Sh_0 accounts for the bulk convection of a non-evaporating spherical droplet. Re_p is the droplet Reynolds number and Sc_p is the Schmidt number while $f(\mathrm{Re}_p)$ is an empirical function defined as $f(\mathrm{Re}_p) = \mathrm{Re}_p{}^{0.77}$ and F_M is a correction factor, which takes into consideration the relative change of the mass film thickness due to the droplet evaporation process:

$$F_M = F_M(B_M) = (1 + B_M)^{0.7} \ln \frac{(1 + B_M)}{B_M}. \tag{7}$$

The evaporation is coupled with the droplet energy/heating in (4) where the droplet temperature variations is described by

$$\frac{dT_p}{dt} = -\frac{6\dot{Q}_l}{\rho \pi c_{pL} D_p^3}, \tag{8}$$

where c_{pL} denotes the heat capacity coefficient of the liquid and $\dot{Q}_l$ the heat flux rate penetrating into the droplet.

The latter is calculated similarly to the evaporation rate and related to it by:

$$\dot{Q}_l = \dot{m}_{p,v} \left(\frac{c_{pm} \left(T_\infty - T_p\right)}{B_T} - h_v(T_P) \right), \tag{9}$$

where $h_v(T_P) = r(T_P) + \dot{Q}_l / \dot{m}_{p,v}$ expresses the effective latent heat of vaporization. It is a function of temperature and varies with the considered liquid. $r(T_P)$ is the true latent heat of vaporization and B_T is the Spalding heat transfer number expressed by

$$B_T = \frac{\overline{c_{p,v}}(T_\infty - T_P)}{r(T_p) + \dot{Q}_l / \dot{m}_{p,v}}. \tag{10}$$

It is related to the mass transfer number B_M by

$$B_T = (1 + B_M)^\phi - 1, \tag{11}$$

where

$$\phi = \frac{c_{p,v}}{c_{p,m}} \frac{Sh^*}{Nu^*} \frac{1}{\overline{\overline{Le}}}. \tag{12}$$

The variable ϕ depends on the thermo-physical properties, the Lewis number Le, and the modified Sherwood and Nusselt numbers (Sh^* and Nu^*). The modified Nusselt number in (12) is defined by:

$$Nu^* = 2 + \frac{Nu_0 - 2}{F_T} \tag{13}$$

where

$$Nu_0 = 1 + (1 + \mathrm{Re}_p Pr)^{1/3} f(\mathrm{Re}_p). \tag{14}$$

Nu_0 expresses the Nusselt number for a non-evaporating spherical droplet with $f(\mathrm{Re}_p)$ as defined above. Pr is the Prandtl number and F_T is a correction factor which accounts for the change of the temperature within the film thickness due to the droplet evaporation process:

$$F_T = F_T(B_T) = (1 + B_T)^{0.7} \ln \frac{(1 + B_T)}{B_T}. \tag{15}$$

In (12) c_{pv} and c_{pm} are the heat capacity coefficients of vapor and mixture, respectively. Between Nu_0 and the modified Nusselt number Nu^* the relationship (13) is considered. No correction to the drag coefficient due to the evaporation process has been taken into consideration. The gas viscosity in the droplet Reynolds number near the droplet has been estimated at a well-defined reference state of temperature and vapor mass fraction according to the averaging "1/3 rule".

2.2 The Non-Equilibrium Evaporation Model

According to [19] and [9] non-equilibrium effects can be captured by determining the departure from thermodynamic equilibrium by adding a deviation term to the molar fraction of the vapor at the droplet surfaces. Thus, the molar mass fraction $\upsilon_{s,eq}$in equation (3) is replaced by $\upsilon_{s,neq}$, for the calculation of vapor mass fraction on the droplet surface. It is determined by the following relation

$$\upsilon_{s,neq} = \upsilon_{s,eq} - \left(\frac{L_K}{d/2}\right) \beta_L \tag{16}$$

where

$$\beta_L = -\left(\frac{3\mathrm{Pr}_G \tau_d}{2}\right) \frac{\dot{m}_{p,v}}{m_p} \tag{17}$$

represents the half of the blowing Péclet number. Pr_G is the Prandtl number, L_K represents the Knudsen length and τ_d is the particle relaxation time.

The Lagrangian equation describing the transient temperature of a single droplet is given by:

$$\frac{dT_p}{dt} = \frac{f_2 Nu}{3\mathrm{Pr}_G} \left(\frac{\theta_1}{\tau_p}\right) (T_G - T_p) + \left(\frac{h_v}{c_{p,l}}\right) \frac{\dot{m}_{p,v}}{m_p} \tag{18}$$

where θ_1 denotes the ratio of the gas heat capacity to that of the liquid phase.

The function f_2 is given by:

$$f_2 = \frac{-\left(\frac{3\mathrm{Pr}_G \tau_p}{2}\right) \frac{\dot{m}_{p,v}}{m_p}}{e^{-\left(\frac{3\mathrm{Pr}_G \tau_d}{2}\right) \frac{\dot{m}_{p,v}}{m}} - 1}. \tag{19}$$

The droplet evaporation influences interaction processes between carrier gas and dispersed phase by means of mass, momentum, temperature, turbulence quantities, and species concentration. These are described according to the Eulerian-Lagrangian methodology adopted in this study and outlined in Sect. 4.

3 Experimental Configurations

Two configurations with two different fuels, i.e. different thermodynamic properties, are investigated. The configurations exhibit completely different boundary conditions and polydisperse turbulent multiphase flows with different classes and probability distribution of the droplet diameters. These are the acetone spray jet representing the phenomena in an open configuration geometry and an evaporating spray in a confined circular chamber. The question is how the suggested models are successful in accurately predicting droplet evaporation in turbulent environments.

3.1 Acetone Spray Jet Configuration

The geometry of the configuration used to study the acetone spray evaporation and/or combustion is shown in Fig. 2. The burner is mounted vertically in a wind tunnel that supplies a co-flowing air stream of 4.5 m/s. The co-flow is provided within a diameter of 104 mm. The spray is initialized 215 mm upstream of the nozzle exit plane. A detailed description of the experimental setup and apparatus used for the generation of the experimental data is given by Masri and Gounder [18] and Stårner et al. [31].

Various cases have been investigated. Table 1 shows the configuration details for various cases investigated, denoted by SP1, SP3, SP6 and SP8. A decreasing mass loading $\dot{m}_l/\dot{m}_{tot}$ in the inner jet is recorded. It equals 33.33, 25 and 19.95 % for the last test cases respectively. The turbulent kinetic energy of the carrier phase is given 5 % of the bulk velocity at the inlet.

The LES are performed using 12 different classes of droplets. The computational domain counts 1.1×10^6 control volumes (cv). The total number of the numerically tracked droplets exceeded 1 millions parcels within one coupling-iteration.

3.2 Evaporating Spray in a Combustor

The second configuration that allows studying the vaporization and mixture of a non-reacting spray is investigated using a RANS method. It consists of a single combustor corresponding to an isopropyl spray issuing into a co-flowing heated air-stream, see Fig. 1. The inner diameter of the main section is 200 mm. The annulus has an outer diameter of 64 mm and an inner diameter of 40 mm. This is represented by a computational domain with $80 \times 39 \times 29$ cells in axial, radial and tangential

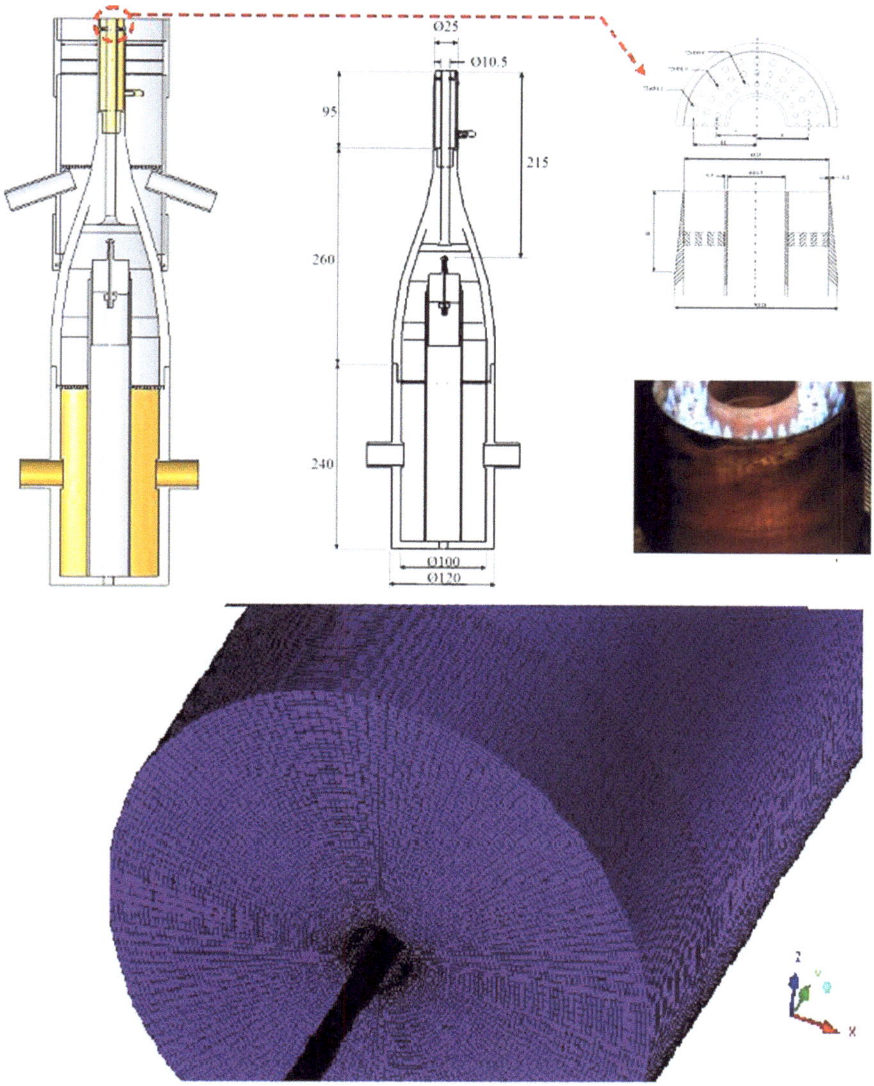

Fig. 2 Configuration of the acetone spray configuration: geometry (*top*), numerical grid (*bottom*)

Table 1 Different test cases with the flow BC

Test case description	SP1	SP3	SP6	SP8
$\dot{m}_l[g/\min]$	75	45	45	45
$U_{jet}[m/s]$	24	24	36	48
$\dot{m}_{air}[g/\min]$	150	150	225	301
$\mathrm{Re}_{jet}[-]$	24,417	20,730	28,076	35,526

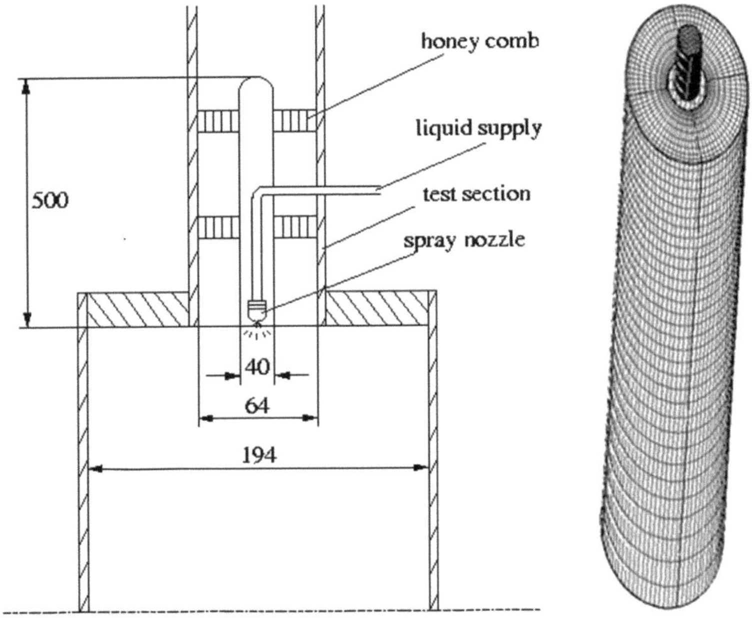

Fig. 3 Spray issuing into a co-flowing heated air-stream: configuration (*left*), grid (*right*)

direction, respectively. The inlet conditions used for the carrier gas correspond to the measured values from the experiment at x = 3 mm downstream. The hollow-cone pressure atomizer in the cylindrical center-body of the inlet tube injects 8 types of droplet classes. These classes are distinguished by the droplet diameter, start velocities, start location and the PDF. These were used as inlet conditions for droplets. The initial temperatures of the droplets and gas are set to 32 °C and 28 °C respectively. The co-flow has a temperature of 80 °C. The maximum air velocity in the considered test case was 18 m/s. The total droplet flow rate was 0.44 g/s (Fig. 3).

4 Numerical Modeling

To account for the instantaneous flow properties encountered by the droplets, involving each particle history starting from the injection into the flow, the Euler-Lagrange approach is adopted. The particles are described by a Lagrangian transport through a continuous carrier gas flow which is captured by an Eulerian approach.

Table 2 Source terms due to the presence of droplets

ψ	$\tilde{S}_{\psi,i}$
1	$\sum_p \frac{\dot{m}_{p,v} N_p}{V_{i,j,k}}$
$\tilde{u}$	$-\sum_p \frac{\dot{m}_p N_p}{V_{i,j,k}} \left[\left(u_p^{t_n+\Delta t} - u_p^{t_n} \right) - g_x \, \Delta t \right] + \sum_p \frac{\dot{m}_{p,v} N_p}{V_{i,j,k}} u_p^{t_n}$
$\tilde{v}$	$-\sum_p \frac{\dot{m}_p N_p}{V_{i,j,k}} \left[\left(v_p^{t_n+\Delta t} - v_p^{t_n} \right) - g_y \, \Delta t \right] + \sum_p \frac{\dot{m}_{p,v} N_p}{V_{i,j,k}} v_p^{t_n}$
$\tilde{w}$	$-\sum_p \frac{\dot{m}_p N_p}{V_{i,j,k}} \left[\left(w_p^{t_n+\Delta t} - w_p^{t_n} \right) - g_z \, \Delta t \right] + \sum_p \frac{\dot{m}_{p,v} N_p}{V_{i,j,k}} w_p^{t_n}$
$\tilde{y}$	$\sum_p \frac{\dot{m}_{p,v} N_p}{V_{i,j,k}}$

4.1 Eulerian Description

The filtered transport equation for the vapour mass fraction is solved together with the filtered transport equations for mass conservation and momentum given in Eqs. (20)–(22) as:

$$\frac{\partial \bar{\rho}}{\partial t} + \frac{\partial \bar{\rho} \tilde{u}_i}{\partial x_i} = \bar{S}_{vapor} \tag{20}$$

$$\frac{\partial}{\partial t}(\bar{\rho}\tilde{u}_i) + \frac{\partial}{\partial x_j}(\bar{\rho}\tilde{u}_i\tilde{u}_j) = -\frac{\partial \bar{p}}{\partial x_i} + \bar{\rho} g_i + \frac{\partial}{\partial x_j}\left[\bar{\rho}\tilde{\nu}\left(\frac{\partial \tilde{u}_i}{\partial x_j} + \frac{\partial \tilde{u}_j}{\partial x_i}\right) - \frac{2}{3}\bar{\rho}\tilde{\nu}\frac{\partial \tilde{u}_k}{\partial x_k}\delta_{ij} - \bar{\rho}\tau_{ij}^{sgs}\right] + \bar{S}_{u,i} \tag{21}$$

$$\frac{\partial}{\partial t}\bar{\rho}\tilde{y} + \frac{\partial}{\partial x_i}(\bar{\rho}\tilde{u}_i\tilde{y}) = \frac{\partial}{\partial x_i}\left(\bar{\rho}\tilde{D}_f\frac{\partial \tilde{y}}{\partial x_i}\right) - \frac{\partial}{\partial x_i}\left(\bar{\rho}J_i^{sgs}\right) + \bar{S}_{vapor} \tag{22}$$

where the dependent filtered variables are obtained from spatial filtering, $\phi = \tilde{\phi} + \phi''$ with $\tilde{\phi} = \overline{\rho\phi}/\bar{\rho}$. Bars and tildes express mean and filtered quantities respectively.

In Eqs. (20)–(22) the variables

u_i (i = 1, 2, 3)	denote the velocity components in x_i direction,
ρ	the density,
p	the hydrostatic pressure and
δ_{ij}	the Kronecker delta.
The quantity ν	is the kinematic molecular viscosity and
D_f	the molecular diffusivity coefficient.

The source terms $\bar{S}_{u,i}$ and $\overline{S}_{vapor}$ that characterize the direct interaction of mass, momentum, and mixture fraction between the droplets and the carrier gas are summarized in Table 2. The variable ψ represents the mean value of mass density, velocity components (u, v, w) and the vapor mass fraction, respectively. The quantities $\tilde{u}$, $\tilde{v}$

and $\tilde{w}$ are the filtered gas phase (axial, tangential and transversal) velocity components while u_p, v_p and w_p represent the three velocity components of the parcel. N_p is the number of real droplets represented by one numerical droplet. $V_{ijk}$is the cell volume. The quantity g represents the gravitation and $\dot{m}_p$ the droplet's mass flow rate across a control volume (CV) per second. Δt stands for the Lagrangian integration time step.

4.2 Lagrangian Description

In addition to the equations for the vaporization in Sect. 2, the position and the trajectory of the droplet need to be tracked. Since the ratio of the density of the disperse phase and that of the carrier phase is approximately three orders of magnitude greater than unity, the forces considered that contribute to the droplet motion are drag, gravitation and buoyancy forces:

$$\frac{du_{p,i}}{dt} = \frac{3}{4}\frac{C_W}{d_p}\frac{\rho}{\rho_p}\left|\vec{u} - \vec{u_p}\right|(u_i - u_{p,i}) + \frac{(\rho_p - \rho)}{\rho_p}g_i = \frac{1}{m_p}\sum_j F_j \tag{23}$$

The drag coefficient C_W used to model the complex dependencies between the particle and the flow conditions is not constant but depends on relative velocities, viscosity of the disperse phase and carrier phase, the particle radius and shape and the roughness of the particle's surface. The drag coefficient used within this work is determined for a spherical, non deformable droplet.

The droplets are tracked until they reach the exit boundary, or are completed vaporized. Average values and variances of droplet characteristic variables (velocity, temperature, etc.) are evaluated in each cell. In case of Large Eddy Simulation (LES), the dispersion of the spray droplets is less relevant since at least 80 % of the instantaneous carrier phase turbulence level should be captured by mesh resolution. However, reports from the literature highlighted the importance of the SGS in the prediction of the disperse phase properties. Bini and Jones [5] mentioned that the SGS dispersion plays a crucial role in determining the spreading and properties of the disperse phase. Dianat et al. [10] showed that the incorporation of a SGS dispersion model tends to improve the predictions of concentration. Shotorban et al. [27] demonstrated through an asymptotic approach that LES may not be able to predict the particle concentration when the SGS effects are neglected on particles in an isotropic turbulent flow. In a first attempt for the sake of simplification, the dispersion of droplet is not accounted for in this study.

To simulate the dispersion of particles and their interaction with the turbulent flow in RANS method, the Markov-sequence model based on the calculation of Lagrangian and Eulerian correlation factors [3] is applied. The particle injection is based on a stochastic approach by considering the particle mass flux and the particle size distributions obtained from the experimental measurements.

The particle sizes are sampled from the particles distribution functions obtained experimentally. The number of injected parcels in one time step equals 2000. The parcels are initialized with predefined velocities given from the experimental measurements, while the injection position is determined randomly with a normal distribution around the axis of the injection hole. Average values and variances of particles characteristic variables are evaluated in each cell.

For the Eulerian description of the turbulent gas phase far from the droplet, the simulation is performed using the three dimensional CFD-code FASTEST in which the equations are solved by finite volume method. The time integration is achieved explicitly with the Runge Kutta method while the diffusion terms are discretized with central schemes on a non-orthogonal block structured grid. The velocity-pressure coupling is accomplished by a SIMPLE algorithm. The whole system is solved by the SIP-solver.

5 Results and Discussion

5.1 Acetone Spray Jet Configuration

The numerical results, presented in this sub-section, are obtained using the non-equilibrium evaporation model. In order to highlight the influences of non-equilibrium conditions and the phase transition behavior of small droplet, a comparison with the equilibrium model is also given. Fist, the validation of the spray dynamics should be assured. Figures 4 and 5 show the axial and radial velocity profiles at different cross sections respectively. The test cases Sp3, Sp6 and Sp8 are compared with the experimental measurements where reasonable agreements are observed. As the eulerian phase is described by means of LES, the gas phase velocity is well captured, making this the source of error that is transmitted to the disperse phase to be significantly reduced. The spray axial velocity-component is an important parameter for the pre-vaporization, since it defines the residence time of every droplet in the computational domain, which in turn influences the vapor distribution in the configuration. The radial component does not demonstrate important values. A small peak is observed at the nozzle exit plane due to the sudden expansion in the geometry.

Figure 6 shows the axial droplet velocity fluctuations at different cross sections. The overall agreement with the experimental data is reasonable. Discrepancies are observed for Sp6 and Sp8 at the centerline, close to the nozzle exit plane. In this region the local droplet volume fraction is larger than 10^{-6} which generally represents the limit for the use of two ways coupling. Important physical phenomena, i.e. four ways coupling, wall interaction, atomization are not modeled. The effect in the nozzle exit area is not investigated, neither experimentally nor numerically. Further downstream, the axial RMS values demonstrate an almost uniform profile. The data in the first and second cross sections are limited to $r/D = 1$ because of the insufficient statistics.

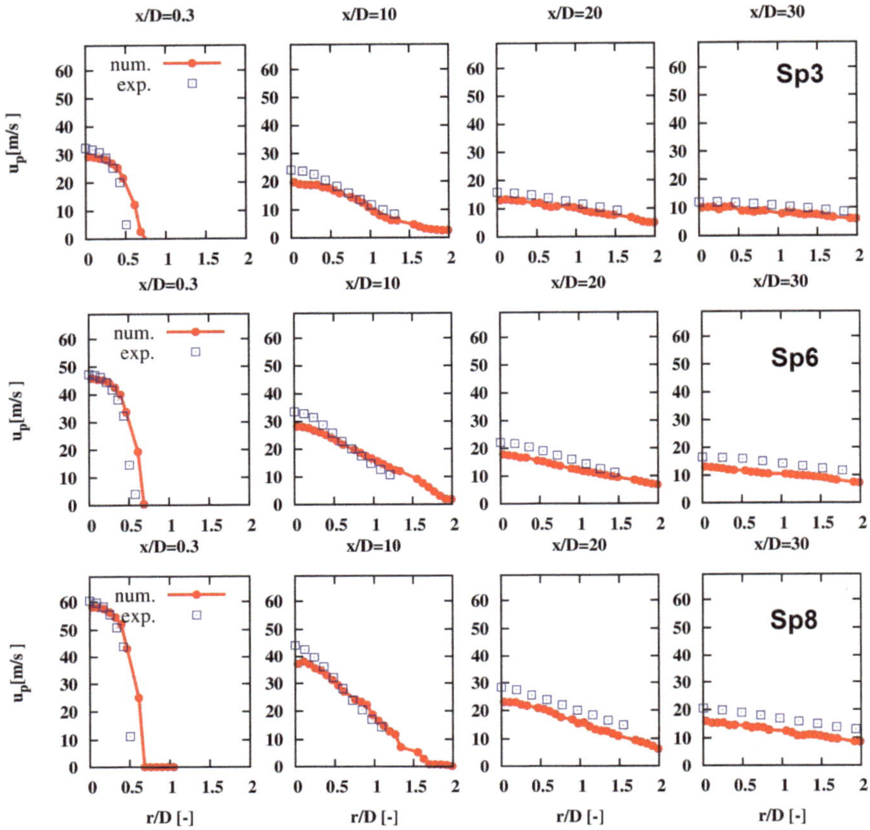

Fig. 4 Axial droplet mean velocity at different cross sections: Comparison between numerical simulation and experimental measurements

In the region r/D > 1 the number of droplets are significantly less in the core of the configuration. This explains the wrinkling at x/D = 10 (Sp8).

The droplet mean diameter and the axial volume flux are presented in Figs. 7 and 8 respectively. In the curve at x/D = 0.03, which depicts the nozzle exit, disagreements are observed for Sp6 and Sp8. The effect of atomization may play an important role in this region. The mean diameter does not seem to be affected significantly by the increase of the carrier phase inlet velocity. The same observation can be made for the axial volume flux. The mixture is likely to have reached the saturation of the fuel concentration, not allowing further evaporation within the 10 mm tube. In the remaining cross sections reasonable agreement is observed.

The profiles of the volume flux and mean diameter, shown in Figs. 7 and 8, delivered a better numerical prediction of the phase transition if compared with the equilibrium model, as it will be discussed in Fig. 9. Important to notice that both

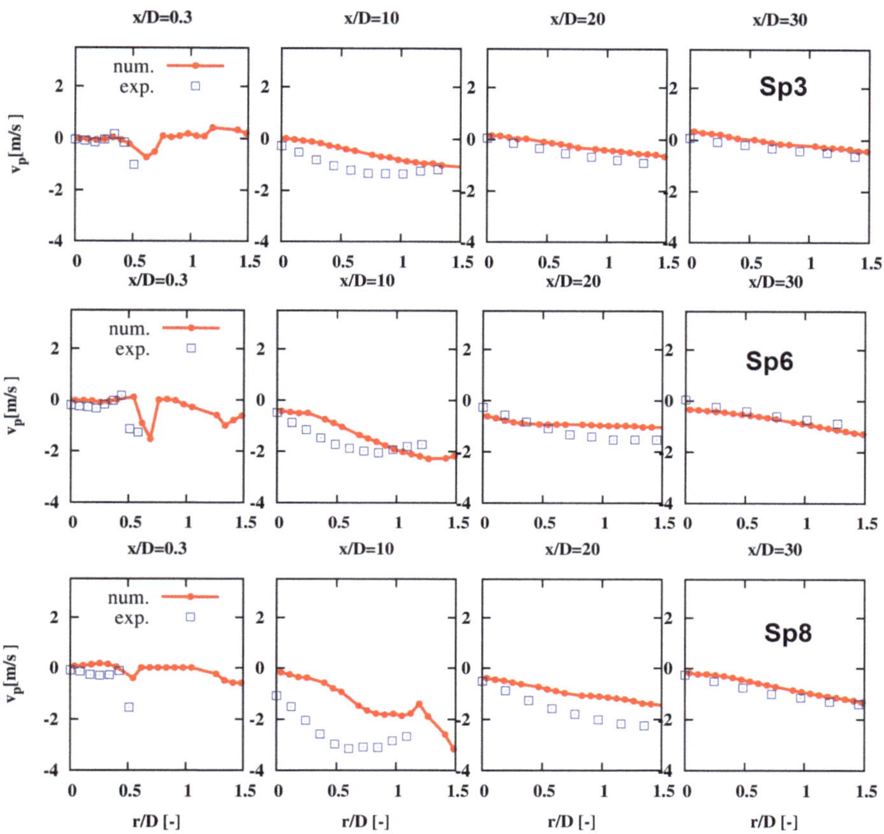

Fig. 5 Radial droplet mean velocity at different cross sections: Comparison between numerical simulation and experimental measurements

model are sensitive to the modeling of the thermodynamic properties, i.e. latent heat, diffusion coefficient, viscosity and the density variation etc. Both models are affected by the droplet injection conditions, i.e. probability density distribution of the droplet classes, injection distribution in time and space and starting values of temperature, velocities with the corresponding RMS values.

In Fig. 9 a comparison of the equilibrium and non-equilibrium vaporization models with the experimental data of the axial volume flux of the spray is shown. The results depict the test case SP1. The volume flux along with the droplets mean diameter is good to quantify the amount of vapor released within each control volume. Both models over-predict the vaporization at the nozzle exit plane. Possible causes for the discrepancies are the interaction with nozzle edges and dense spray which may necessitate four-way coupling. In the remaining cross section downstream the nozzle exit, the non-equilibrium model results agree reasonably with the measurements. The

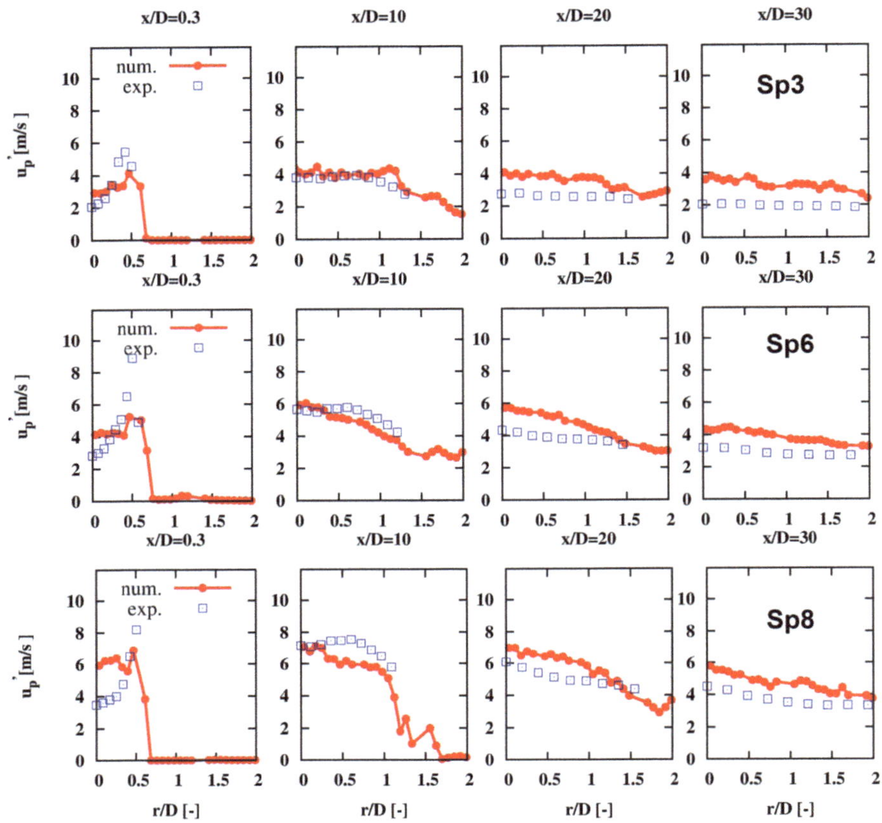

Fig. 6 Root Mean Square of the droplet axial velocity fluctuations at different cross sections. Comparison between numerical simulation and experimental measurements

effects of non equilibrium are likely to be present in the two phase flow as the droplet mean diameter is between 20 and 40 μm.

5.2 *Evaporating Spray in a Combustor*

Details of the velocity field and turbulence quantities are given in [9]. Figure 10 shows the radial distribution of the droplet mass flux for the second configuration at different axial positions. In accordance with the experiment, the concentration of droplets decreases while moving away from the nozzle due to the evaporation. A comparison between equilibrium and non-equilibrium evaporation models reveals that the latter delivers results closer to experimental results. The evaporation rate predicted by the equilibrium model is nearly constant. The increase of surface to radius ratio is not remarkably affecting the phase transition. I.e. the equilibrium

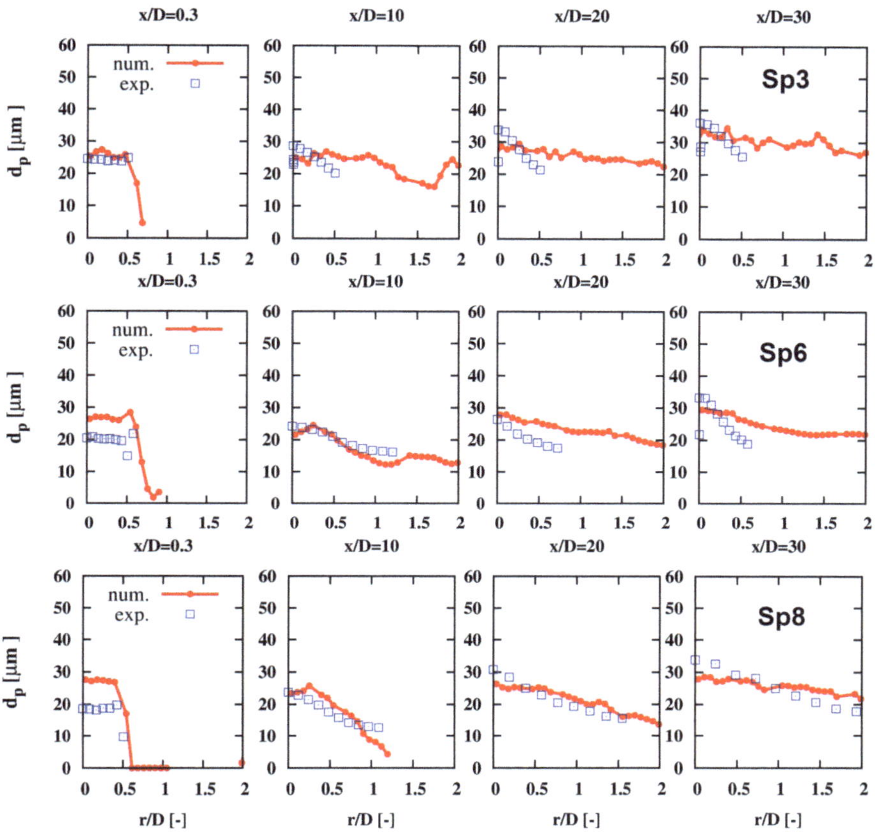

Fig. 7 Droplet mean diameters of the evaporating spray

model is not able to properly predict the correct droplet mass flux. As the spray is located at the center of the configuration, the effect of confinement is not obviously observed. The confinement increases the shear flow at the wall boundaries and may produce secondary recirculation zones at the corners of the configuration. This would increase the droplet residence time in the regions close to the injection nozzle and thus improve the evaporation rate. The spray, in the configuration under study, is bounded within the radius of 40 mm, whereas the outer diameter is 194 mm. The shear flow at the wall is not important for the droplets at the center of the configuration.

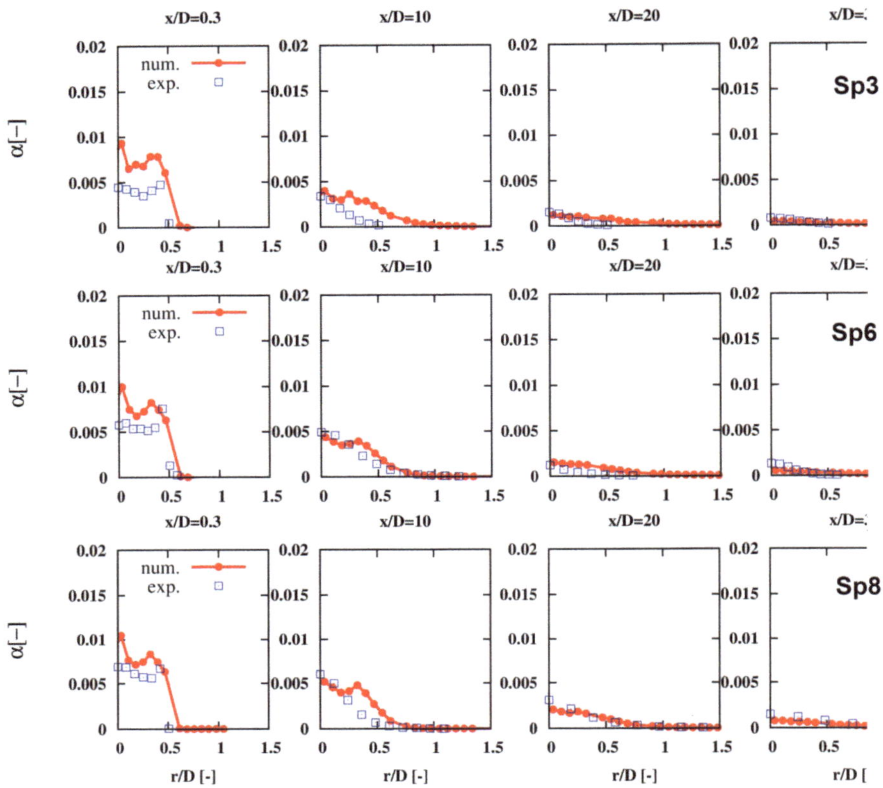

Fig. 8 Spray axial volume flux in [m3/(m2.s)]. Numerical results are obtained using the non-equilibrium vaporization model

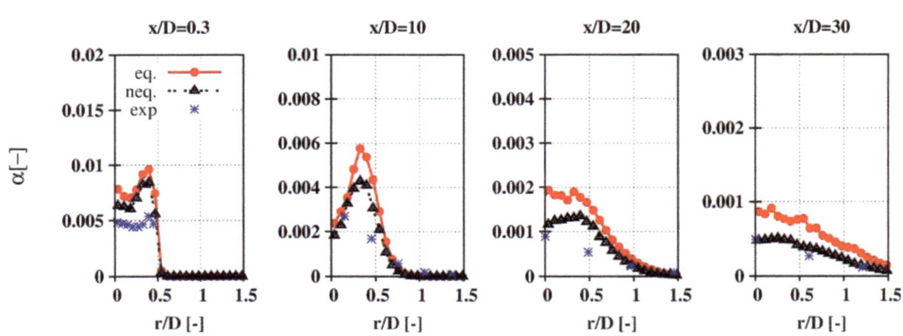

Fig. 9 Prediction of the spray axial volume flux in [m3/(m2.s)] of the acetone test SP1. Comparison of the equilibrium and non-equilibrium vaporization models with the experimental results

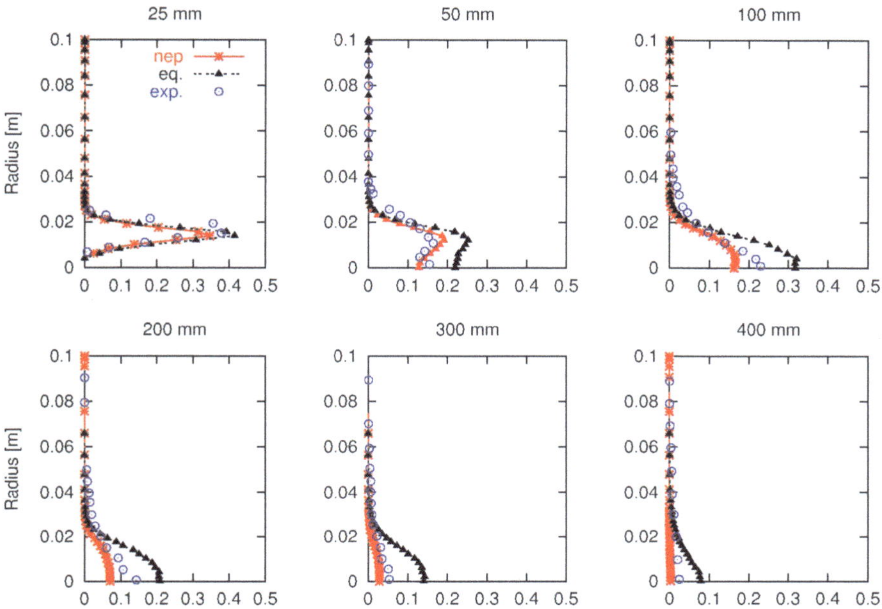

Fig. 10 Droplet mass flux: comparison between results with equilibrium and non equilibrium evaporation models

6 Conclusions

To capture droplet evaporation in a reasonable accuracy with economical computational costs, the Uniform Temperature Model (UTM) by Abramzon and Sirignano for equilibrium evaporation is recommended. Since in spray flame modeling non-equilibrium processes are usually present (especially close to the combustion zone and also at the end of a droplet life), the UTM should be extended to capture non equilibrium effects. The model by Langmuir-Knudsen accomplishes good agreement with experimental data. At the same time the computational costs do not increase considerably if compared to equilibrium simulation. Since simulations are done usually in poly-disperse flows, spray should be represented using as many classes as possible, since evaporation rates differ from one droplet diameter to another.

References

1. Bellan J., Harstad K., 1987: Analysis of the convective evaporation of nondilute clusters of drops. Int. J. Heat Mass Transfer 30 (1),125–136.
2. Bellan J. and Selle L. C.: “Large Eddy Simulation composition equations for single-phase and two-phase fully multicomponent flows” Proceedings of the Combustion Institute Volume 32,

Issue 2, Pages 2239–2246, 2009.
3. Berlemont A., Grancher M. S., Gouesbet G.: On the Lagrangian simulation of turbulence influence on droplet evaporation. Int. J. Heat Mass Transfer 34 (11) 2805–2812, 1991.
4. Berlemont A., Grancher M. S. and Gouesbet G.: Heat and mass transfer coupling between vaporizing droplets and turbulence using a Lagrangian approach. J. of. Heat and Mass Transfer Vol.38 (1995) 3023–3034.
5. Bini M. and Jones W. P., Large-eddy simulation of particle-laden turbulent flows, J. Fluid Mech. vol. 614, 2008, pp. 207–252.
6. Bini M., Jones W. P. "Large Eddy Simulation of an evaporating acetone spray" International Journal of Heat and Fluid Flow, Volume 30, Issue 3, pp. 471–480, 2009.
7. Burger M., Schmehl R., Prommersberger K., Sch€ afer O., Koch R., Wittig S.: "Droplet evaporation modeling by the distillation curve model: accounting for kerosene fuel and elevated pressures" International Journal of Heat and Mass Transfer 46 4403–4412, (2003).
8. Chrigui M. and Sadiki A.: "Effects of Turbulence Modulation on Mass and Heat Transfer: 3D-Numrical Prediction Based on Coupled Advanced Models for Turbulence and Evaporation" 4th Int. Symp. On Turbulence, Heat and Mass Transfer, October 12–17, Antalya, Turkey.
9. Chrigui, M., Ahmadi G., Sadiki A., "Study on Interaction in Spray between Evaporating Droplets and Turbulence Using Second Order Turbulence RANS Models and a Lagrangian Approach", Progress in Computational Fluid Dynamics, Special issue, 2004.
10. Dianat M., Yang Z. and McGuirk J. J. Large-Eddy Simulation of a Two-Phase Plane Mixing-Layer, Advances in Turbulence XII, Springer Proceedings in Physics, 2009, Volume 132, Part 11, 2009, 775–778.
11. Faeth G. M.: Evaporation and combustion of sprays: Progress in energy and combustion science, Pergamon Press, 9:1–76, 1983.
12. Gökalp I., Chauveau C., Simon O., Chesneau X., "Mass transfer from Liquid fuel droplets in turbulent flow", The combustion Institute 1992.
13. Hallett W. L. H.: "A simple model for the vaporization of droplets with large numbers of components", Combustion and Flame 121, 334–344, 2000.
14. Kohnen, G., Rüger M. and Sommerfeld M. "Convergence behavior for numerical calculations by the EULER/LAGRANGE Method for strongly coupled phases." FED-Vol 185, Numerical Methods in Multiphases Flows ASME 1994.
15. Law C. K.: Unsteady droplet vaporization with droplet heating. Combustion and Flame, 26: 17–22, 1976.
16. Lederlin T. and Pitsch H.: "Large-eddy simulation of an evaporating and reacting spray" Center for Turbulence Research Annual Research Briefs, 2008.
17. Maqua C., Castanet G., Lemoine F.: "Bicomponent droplets evaporation: Temperature measurements and modelling" Fuel, Volume 87, Issues 13–14, pp. 2932–2942, 2008.
18. Masri A. R. and Gounder J. D. Proc. of the 6th Mediterranean Com. Symp, 2009.
19. Miller R. S., Harstad K., Bellan J.: Evaluation of equilibrium and non-equilibrium evaporation models for many droplet gas liquid flow simulation, Int. J. of Multiphase flow, 24, pp. 1025–1055, 1998.
20. Oefelein J. C., Aggarwal S. K.: Toward a unified high pressure drop model for spray simulation, center for turbulence research, Proceedings of summer program 2000, pp. 193–205.
21. Pera C., Réveillon J., Vervisch L., Domingo P. "Modeling subgrid scale mixture fraction variance in LES of evaporating spray Combustion and Flame", Volume 146, Issue 4, pp. 635–648, 2006.
22. Prommersberger K., Maier G., and Wittig S. (1998). Validation and Application of a Droplet Evaporation Model for Real Aviation Fuel, Proceedings of NATO-RTO Meeting on Gas Turbine Engine Combustion, Emissions and Alternative Fuels, Lisbon, Portugal, 12–16 October.
23. Prommersberger K., Stengele J., Dullenkopf K., Himmelsbach J. and Wittig S. "Investigations of droplet evaporation at elevated pressures" Collaborative Research Center 167, September 1998 Karlsruhe, Germany.
24. Sanjosé M., Lederlin T., Gicquel L., Cuenot B., Pitsch H., García-Rosa N., Lecourt R. and Poinsot T.: "LES of two-phase reacting flows", Center for Turbulence Research Proceedings of the Summer Program, 2008.

25. Selle L. C. and Bellan J.: "Characteristics of transitional multicomponent gaseous and drop-laden mixing layers from Direct Numerical Simulation: Composition effects", Physics of Fluids, 19, 063301-1-33, 2007.
26. Senoner J. M., Sanjosé M., Lederlin T., Jaegle F., García M., Riber E., Cuenot B., Gicquel L., Pitsch H., Poinsot T. "Eulerian and Lagrangian Large-Eddy Simulations of an evaporating two-phase flow" Comptes Rendus Mécanique, 2009.
27. Shotorban B., Zhang K. K. Q., Mashayek F. Improvement of particle concentration prediction in large-eddy simulation by defiltering, International Journal of Heat and Mass Transfer, Volume 50, Issues 19–20, 2007, Pages 3728–3739.
28. Sirignano W. A. "Fluid dynamics of sprays" 1992 Freeman Scholar Lecture J. Fluids Engng. Vol. 115, pp. 345–378. 1993.
29. Sirignano W. A., Wu G. "Multicomponent-liquid–fuel vaporization with complex configuration", International Journal of Heat and Mass Transfer, Volume 51, Issues 19–20, pp. 4759–4774, 2008.
30. Sornek, R. J., Dobashi R., Hirano T., "Effect of Turbulence on Vaporization, Mixing, and Combustion of Liquid-fuel Sprays", Combustion and Flame 120:479–491, 2000.
31. Stårner S. H., Gounder J., and Masri A. R., Com. & Flame, Vol. 143, pp. 420–432, Dec. 2005.
32. Zugasti M. A., Rosner D. E. "Multicomponent fuel droplet vaporization and combustion using spectral theory for a continuous mixture" Combustion and Flame, Volume 135, Issue 3, Pages 271–284, 2003.

Theoretical and Numerical Study of Evaporation Effects in Spray Flamelet Modeling

Hernan Olguin and Eva Gutheil

Abstract Even though laminar spray flames differ considerably from their gaseous counterpart, most often flamelet models employed in the simulation of turbulent spray combustion are based on laminar gas flame structures neglecting the influence of spray evaporation in the laminar spray flamelet. In this work, a combined theoretical and numerical study of the impact of spray evaporation on the structure of laminar spray flames is presented. Numerical simulations of an axisymmetric laminar monodisperse ethanol/air counterflow spray flame are performed in order to evaluate the influence of spray evaporation on flame characteristics. Flame structures for different initial droplet radii and strain rates are considered. Special emphasis is given to the effects of the spray on extinction and on different local combustion regimes. Moreover, the classical flamelet equations are reinvestigated, and the derivation of spray flamelet equations is presented, where additional terms caused by spray evaporation are identified—the classical gas flamelet equations are recovered for non-evaporating conditions. Two new terms accounting for evaporation and for combined mixing and evaporation, respectively, are identified, and their relative importance is presented and discussed for the numerical spray flame structures. The results show that the distribution of the spray evaporation rate plays a key role in the characterization of the spray flame structure. The new source terms overweigh the dissipation term of the gas phase in most situations and regimes of the flame even for non-evaporating species. Therefore, spray evaporation should always be considered. The relevance of the present formulation for turbulent spray modeling is evaluated and discussed, and a novel spray flamelet formulation is suggested.

H. Olguin (✉) · E. Gutheil
Interdisciplinary Center for Scientific Computing, University of Heidelberg,
Im Neuenheimer Feld 368, 69120 Heidelberg, Germany
e-mail: hernan.olguin@iwr.uni-heidelberg.de

E. Gutheil
e-mail: gutheil@iwr.uni-heidelberg.de

B. Merci, E. Gutheil (eds.), *Experiments and Numerical Simulations of Turbulent Combustion of Diluted Sprays*, ERCOFTAC Series 19, DOI 10.1007/978-3-319-04678-5_4,
© Springer International Publishing Switzerland 2014

1 Introduction

Turbulent spray combustion is relevant in many technical applications such as internal combustion engines, industrial furnaces and household burners, turbines and liquid-fueled rocket engines. The consideration of detailed chemical reaction mechanisms in the simulation of these processes is of vital importance for the investigation and reduction of pollutant emission during the operation of these devices. Unfortunately, a direct inclusion of detailed chemical reactions dramatically increases the computational cost of the numerical simulations of technical combustion processes, and it is prohibitive in practical situations.

Flamelet models have been derived in order to include detailed chemical reactions in turbulent gas flame simulations [10, 24–26]. In this approach, a turbulent flame is considered to be an ensemble of laminar flamelets [24]. This assumption is valid for high Damköhler numbers, which is fulfilled in many technical combustion applications. The success of flamelet based models in the simulation of turbulent gas flames has motivated their application in the simulation of turbulent spray flames [6, 16]. However, the classical flamelet model includes very strict assumptions regarding the different combustion regimes that may occur in turbulent flames. In general, either non-premixed or premixed combustion are considered, whereas the partially premixed regime is discarded. This issue will be discussed in the present study.

Recently, several multi-regime flamelet models have been developed in order to overcome the limitations related with classical flamelet models [7, 18, 22]. In this context, Franzelli et al. [7] proposed the use of partially-premixed pure gas flame libraries for the simulation of spray flames. In general, multi-regime flamelet models based on laminar gas structures are able to predict the flame characteristics in zones where no evaporation occurs, since these regions are not considerably affected by spray processes [13], but they are not suitable to properly describe the flame structure in zones where both evaporation and combustion occur, since they are dominated by evaporation effects. Therefore, a separation of the regimes with pure gas combustion, i.e. all droplets have vaporized, and a regime where both evaporation and combustion occur simultaneously [13], solves the question of the pure gas combustion regime, but not the region where both evaporation and chemical reactions occur simultaneously [17].

Hollmann and Gutheil [17] and Gutheil [13] proposed an extension of the classical non-premixed flamelet model [24] for spray flames, which consistently uses a library based on laminar spray structures. It is found that spray flamelets are not only determined by the mixture fraction and its scalar dissipation rate (associated with the strain rate) as in counterflowing laminar gas diffusion flames [24], but they also depend on the initial droplet size and velocity and the equivalence ratio on the spray side of the configuration [5, 13, 14, 17]. The inclusion of these additional parameters is required in order to characterize laminar spray flamelets. This has been successfully achieved for turbulent methanol/air [8, 17] and ethanol/air [9] spray flames. However, this approach includes a high-dimensional flamelet library of the order five for the mixture fraction, its scalar dissipation rate, the equivalence ratio, and the initial droplet size and velocity.

Because of the complexity of the formulation, currently, most studies of turbulent spray flames use laminar flamelet structures, where pure gas combustion is considered [6, 16], even though the relevance of evaporation on flamelet structures is known [2, 3, 11, 13, 17, 19–21].

The scope of the present study is the reconsideration of the classical flamelet equations for use in spray combustion with emphasis on the evaporation and its effect on the spray combustion process. For this purpose, the spray flamelet equations are derived and the different contributions are analyzed for laminar ethanol/air spray flames. The new flamelet equations and their evaluation lead to a revised formulation of the flamelet model for spray combustion. The applicability of the present formulation is emphasized in connection with different regimes in laminar spray combustion, which are characterized through analysis of the mixture fraction profile, and regimes of premixed, non-premixed, and partially premixed regimes are identified.

2 Mathematical Model

In this section, the general governing equations are presented, which provide the base for the derivation of the spray flamelet equations [23]. Moreover, the formulation of the equations solved to study the detailed spray flame structures is provided and discussed.

2.1 Governing Equations

The present formulation employs a Eulerian-Lagrangian formulation of a dilute, reactive spray flow in the counterflow configuration. The mono-disperse spray is characterized through equations accounting for droplet heating, evaporation, and motion. A similarity analysis is applied, which transfers the problem into a one-dimensional formulation [5, 14]. This transformation and the formulation for viscous flows allow for the inclusion of detailed chemical reaction mechanisms [12–14].

In this section, the general governing equations for the gas and liquid phase are presented, which are used for the derivation of novel spray flamelet equations.

2.1.1 Gas Phase

The conservation equations of the mass, momentum, mass fractions of chemical species and energy can be written as

$$\frac{\partial \rho}{\partial t} + \frac{\partial (\rho u_i)}{\partial x_i} = S_v \tag{1}$$

$$\rho \frac{\partial u_j}{\partial t} + \rho u_i \frac{\partial u_j}{\partial x_i} = -\frac{\partial p}{\partial x_j} - \frac{\partial \tau_{ij}}{\partial x_i} - u_j S_v + S_{m,j} \tag{2}$$

$$\rho\frac{\partial Y_k}{\partial t} + \rho u_i\frac{\partial Y_k}{\partial x_i} + \frac{\partial}{\partial x_i}\left(\rho V_{k,i} Y_k\right) = \dot{\omega}_k + (\delta_{Fk} - Y_k)\, S_v, \quad k = 1, \ldots, N \tag{3}$$

$$\rho C_p\frac{\partial T}{\partial t} + \rho u_i C_p\frac{\partial T}{\partial x_i} = -\sum_{k=1}^{N} h_k\dot{\omega}_k + \frac{\partial p}{\partial t} + u_i\frac{\partial p}{\partial x_i} + \frac{\partial}{\partial x_i}\left(\lambda\frac{\partial T}{\partial x_i}\right)$$
$$-\rho\frac{\partial T}{\partial x_i}\sum_{k=1}^{N} C_{p,k} Y_k V_{k,i} - \tau_{ij}\frac{\partial u_i}{\partial x_j} - S_v\int_{T_0}^{T} C_{p,F}\mathrm{d}T + S_e, \tag{4}$$

where u_i is the gas velocity in i direction, ρ is the gas density, Y_k is the mass fraction of species k, C_p and $C_{p,k}$ are the specific heat capacity at constant pressure of the mixture and of species k, respectively, p is the static pressure, h_k is the enthalpy of species k and λ denotes the heat conductivity. S_v, $\mathbf{S}_m$, and S_e are sources of mass, momentum and energy, respectively, $V_{i,k}$ is the diffusion velocity, V_k, of species k in i direction into the mixture. δ is the Kronecker symbol, F is fuel and $\dot{\omega}_k$ is the specific chemical reaction rate of species k for $k = 1, \ldots, N$. The viscous tensor τ_{ij} is defined by

$$\tau_{ij} = -\eta\left(\frac{\partial u_i}{\partial x_j} + \frac{\partial u_j}{\partial x_i}\right) + \frac{2}{3}\eta\frac{\partial u_k}{\partial x_k}\delta_{ij} \tag{5}$$

neglecting the bulk viscosity, and η is the dynamic gas viscosity. The terms including the spray mass source, S_v, appearing in Eqs. (2) through (4) are introduced through use of the continuity Eq. (1) in these equations.

The formulation of the spray flamelet equations requires the definition of the mixture fraction. In gas combustion processes including hydrocarbons or alcohols, most often the mixture fraction definition is based on the chemical element C, because this formulation fulfills the requirements of monotonicity and boundedness between zero and unity [15]. In general, the definition of the mixture fraction based on a chemical element, A, yields

$$\xi_{\mathrm{A}} = \frac{Z_{\mathrm{A}} - Z_{\mathrm{A,min}}}{Z_{\mathrm{A,max}} - Z_{\mathrm{A,min}}}, \tag{6}$$

where Z_{A} is the mass fraction of element A and N is the total number of chemical species in the system. Z_{A} can be expressed as

$$Z_{\mathrm{A}} = \sum_{k=1}^{N}\left(\frac{a_{k\mathrm{A}} M_{\mathrm{A}}}{M_k}\right) Y_k, \tag{7}$$

where $a_{k\mathrm{A}}$ denotes the number of moles of element A in species k, and M_{A} and M_k denote the molecular weights of element A and of species k, respectively. In the remainder of the present chapter, the mixture fraction is noted as ξ. Adopting Fick's diffusion law and neglecting thermal diffusion, Eq. (3) can be written as

$$\frac{\partial(\rho Y_k)}{\partial t} + \frac{\partial\left(\rho u_i Y_k\right)}{\partial x_i} - \frac{\partial}{\partial x_i}\left(\rho D_k\frac{\partial Y_k}{\partial x_i}\right) = \dot{\omega}_k + \delta_{Fk} S_v. \tag{8}$$

Multiplication of Eq. (8) by $\frac{a_{k\mathrm{A}} M_{\mathrm{A}}}{M_k}$ and summation over all species N yields the transport equation of the mass fraction of element A, Z_{A}, and with the assumption of equal diffusion coefficients, $D_k = D$, $k = 1, \ldots, N$ for all species k, and following Eq. (6), the transport equation of the mixture fraction, ξ, is obtained as

$$\frac{\partial(\rho\xi)}{\partial t} + \frac{\partial(\rho u_i \xi)}{\partial x_i} - \frac{\partial}{\partial x_i}\left(\rho D \frac{\partial \xi}{\partial x_i}\right) = S_v. \tag{9}$$

In contrast to the derivation of the mixture fraction for gas flames, the present formulation for spray flows includes the source term, S_v, accounting for spray evaporation.

2.1.2 Liquid Phase

Assuming the droplet velocity to evolve by interactive drag applied by the surrounding gas and gravity, the following relation is used to express the equation of droplet motion

$$m \frac{d\mathbf{v}}{dt} = \pi R^2 \frac{1}{2} \rho \left(\mathbf{u} - \mathbf{v}\right) \mid \mathbf{u} - \mathbf{v} \mid C_D + m\mathbf{g}. \tag{10}$$

Here, C_D is the drag coefficient, $\mathbf{v}$ and $\mathbf{u}$ denote the droplet and gas velocity, respectively, R is the instantaneous droplet radius, and $m = \frac{4}{3}\pi R^3 \rho_l$ denotes the mass of a droplet with radius R, and ρ_l denotes liquid density. Droplet evaporation is accounted for by considering the rate of change of droplet mass [1]

$$\dot{m} = 2\pi R \rho_f D_f \widetilde{\mathrm{Sh}} \ln(1 + B_M). \tag{11}$$

The subscript f refers to properties in the film, $\widetilde{\mathrm{Sh}}$ is the modified Sherwood number, $\dot{m}$ is the droplet mass vaporization rate, and B_M is the Spalding mass transfer number, which can be expressed as

$$B_M = \frac{Y_{Fs} - Y_F}{1 - Y_{Fs}}, \tag{12}$$

where Y_{Fs} and Y_F are mass fractions of the fuel vapor at the droplet surface and in the bulk of surrounding gas, respectively.

Droplet heating is modeled through the conduction limit model, which resolves the droplet interior in space. This model is known to be the most accurate for heat conduction within the droplet. Thus, the energy equation for a droplet can be written as

$$\frac{\partial T_l}{\partial t} = \alpha_l \frac{1}{r^2} \frac{\partial}{\partial r}\left(r^2 \frac{\partial T_l}{\partial r}\right), \tag{13}$$

where T_l is the temperature of the liquid and r is the radial coordinate inside the spherical droplet. Finally, when no droplet interaction is considered, the equation for the droplet number density, n, yields

$$\frac{\partial n}{\partial t} + \frac{\partial \left(n\, v_i\right)}{\partial x_i} = 0. \tag{14}$$

Here, v_i denotes the droplet velocity in i direction. For a mono-disperse spray, the spray source terms to enter Eqs. (1) through (4) yield

$$S_v = n\dot{m} \tag{15}$$

$$\mathbf{S}_m = -nm\frac{\mathrm{d}\mathbf{v}}{\mathrm{d}t} + n\dot{m}\mathbf{v} \tag{16}$$

$$S_e = -n\left(\dot{q} + L_V\right) + n\dot{m}\int_{T_0}^{T_s} C_{p,F}\mathrm{d}T, \tag{17}$$

where $\dot{q} = \dot{m}\left[C_{p,F}(T - T_s)/B_T - L_V\right]$ is the energy transferred to the droplet and L_V is the temperature dependent latent heat of vaporization.

For the numerical simulation of laminar counterflowing spray flames, which can be presented in two-dimensional physical space, the similarity transformation derived by Continillo and Sirignano [5] and generalized by Gutheil and Sirignano [14] is used, and non-dimensional equations are solved. The transformed equations for mono-disperse sprays are given in Refs. [5, 14] and for poly-disperse equations in Ref. [12]. The numerical solution procedure is given in Refs. [5, 14].

The present computations concern an ethanol/air spray flow, where a chemical reaction mechanism consisting of 38 species and 337 elementary reactions is implemented [4, 12].

2.2 *Derivation of Spray Flamelet Equations*

The gas flamelet equations derived by Peters [24] are appropiate for pure gas combustion, but they need reconsideration for spray flames. In the present section, the derivation of spray flamelet equations is presented, which account for the evaporation of the droplets. It is assumed that the flamelet structure varies with the local mixture state characterized by the mixture fraction. The aim of this work is the formulation of a flamelet model for spray flames, which incorporates the effect of evaporation on the laminar counterflow spray flame.

The combination of Eqs. (1) and (8) yields

$$\rho\frac{\partial Y_k}{\partial t} + \rho u_i\frac{\partial Y_k}{\partial x_i} - \frac{\partial}{\partial x_i}\left(\rho D_k\frac{\partial Y_k}{\partial x_i}\right) = \dot{\omega}_k + S_v\left(\delta_{Fk} - Y_k\right). \tag{18}$$

Assuming that combustion takes place in a very thin layer, i.e. flamelet, in the vicinity of the surface of stoichiometric mixture fraction, ξ_{st},

$$\xi(x_i, t) = \xi_{\mathrm{st}} \tag{19}$$

and that the structure of this flamelet varies only in a direction normal to this surface, a coordinate system based on the mixture fraction ξ is introduced, where all scalar variables can be expressed as

$$Y_k = f_k\left(\xi, \tau\right). \tag{20}$$

Thus, f_k is a function of the mixture fraction and the transformed time τ. Following the method suggested by Peters [24], Eq. (18) can be transformed from physical space (x_1, x_2, x_3, t) into a new system of coordinates in the mixture fraction space (ξ, τ). Application of the transformation rules

$$\frac{\partial}{\partial t} = \frac{\partial}{\partial \tau} + \frac{\partial}{\partial \xi}\frac{\partial \xi}{\partial t} \quad \text{and} \quad \frac{\partial}{\partial x_i} = \frac{\partial}{\partial \xi}\frac{\partial \xi}{\partial x_i} \tag{21}$$

transforms the species transport equations into

$$\begin{aligned} \rho\frac{\partial Y_k}{\partial \tau} &= \rho D_k\left(\frac{\partial \xi}{\partial x_i}\right)^2 \frac{\partial^2 Y_k}{\partial \xi^2} + \dot{\omega}_k \\ &- \left(\rho\frac{\partial \xi}{\partial t} + \rho u_i \frac{\partial \xi}{\partial x_i} - \frac{\partial}{\partial x_i}\left(\rho D_k \frac{\partial \xi}{\partial x_i}\right)\right)\frac{\partial Y_k}{\partial \xi} + S_v\left(\delta_{Fk} - Y_k\right). \end{aligned} \tag{22}$$

Assuming equal diffusion coefficients for all species and introducing the definition of the scalar dissipation rate, χ, of the mixture fraction

$$\chi = 2D\left(\frac{\partial \xi}{\partial x_i}\right)^2, \tag{23}$$

Eq. (22) can be rewritten as

$$\rho\frac{\partial Y_k}{\partial \tau} = \underbrace{\rho\frac{\chi}{2}\frac{\partial^2 Y_k}{\partial \xi^2}}_{\text{Dissipation}} + \dot{\omega}_k + \underbrace{S_v\left(\xi - 1\right)\frac{\partial Y_k}{\partial \xi}}_{\text{Mixing/Evaporation}} + \underbrace{S_v\left(\delta_{Fk} - Y_k\right)}_{\text{Evaporation}}, \tag{24}$$

where Eq. (9) has been introduced using the product law and Eq. (1).

Equation (24) presents the flamelet equation for laminar spray flames. The first three terms are identical to the flamelet equation for gas flames, whereas the last two terms are new, and they represent the effect of the spray evaporation on the flame structure. Thus, the flamelet formulation for spray flames does not only depend on the mixture fraction and its scalar dissipation rate, but additionally on the spray evaporation source term, S_v.

Considering the consequence for the spray flamelet formulation for turbulent spray flames, this means that not only the turbulent mixing and the scalar dissipation rate affect the flamelets, but evaporation must be taken into account. The previous formulation of Hollmann and Gutheil [17] given as

$$\tilde{\phi} = \int_0^\infty \int_0^\infty \int_0^\infty \int_0^\infty \int_0^1 \phi \tilde{P}\left(\xi, \chi, E, R_0, v_0\right) \mathrm{d}\xi \mathrm{d}\chi \mathrm{d}E \mathrm{d}R_0 \mathrm{d}v_0, \tag{25}$$

may now be replaced by the formulation [23]

$$\tilde{\phi} = \int_{-\infty}^\infty \int_0^\infty \int_0^1 \phi \tilde{P}\left(\xi, \chi, S_v\right) \mathrm{d}\xi \mathrm{d}\chi \mathrm{d}S_v, \tag{26}$$

where $\tilde{\phi}$ denotes the Favre-averaged value of a scalar variable, for instance, the mass fraction of any chemical species in the turbulent spray flame, and ϕ is the corresponding scalar variable in the laminar spray flame.

The new formulation yields a lower dimensional dependence of the physical variables, which requires a lower dimensional probability density function, whereas the quantities equivalence ratio, initial droplet size and velocity might be more straight forward parameters to be determined for the flamelet computations. The major difference between both formulations is that the former formulation requires consideration of input parameters into the laminar flame computation, whereas the evaporation rate in the new formulation is a result from the laminar flame simulation. The question of how to introduce the new formulation into a computer code for turbulent spray flame simulations will be discussed in future.

The present study concerns the evaluation of the influence of the new evaporation terms appearing in Eq. (24). For this purpose, numerical simulations of laminar ethanol/air spray flames are performed, and the flamelet equation for spray flames will be analyzed and discussed.

3 Results and Discussion

This section presents numerical results for ethanol/air spray flames. Earlier studies [12] of ethanol/air spray flames concerned the structure of these flames for bi-disperse spray flames. The present results address the evaluation of evaporation effects and the study of different combustion regimes in these flames. First, evaporation effects on the flame structure are studied in physical space at low and extinction strain rates. Different combustion regimes are identified in spray flames, and the novel flamelet formulation is discussed in this framework.

3.1 Evaporation Effects in Physical Space

The structure of laminar (spray) flames strongly depends on strain. At low strain rate, the spray penetrates into the flame front, and as strain rate is increased, the droplets may cross the flame front and oscillate around the stagnation plane. The droplets re-entering the flame zone strongly enhance evaporation and combustion before flame extinction occurs [14].

In the present numerical computations, liquid ethanol with carrier gas air is injected from the left side of the counterflow configuration, and it is directed against air. Both air and liquid fuel are injected at 300 K at atmospheric pressure. The equivalence ratio, E, is unity for all situations under investigation.

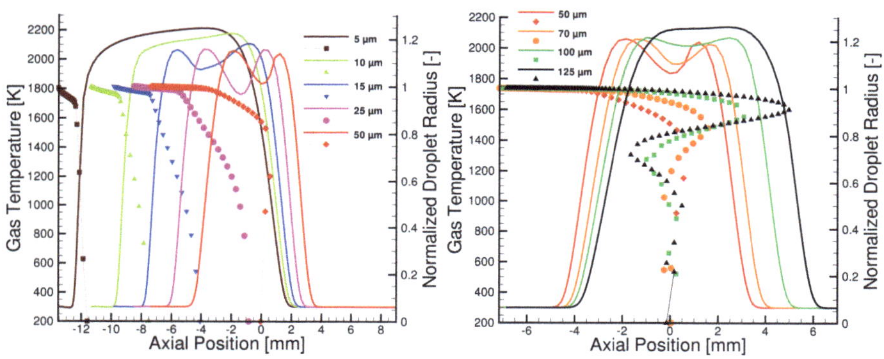

Fig. 1 Gas temperature and normalized droplet radius for initial droplet radii from 5 to 50 μm (*left*) and from 50 to 125 μm (*right*), $a = 55/\mathrm{s}$

3.1.1 Evaporation Effects at Low Strain Rate

For the present computations, the gas strain rate on the spray side of the configuration is $a = 55/\mathrm{s}$ and the initial spray velocity is 0.44 m/s at this strain rate. A grid of 250 nodes is used for all the computations in this subsection.

Figure 1 gives a survey of the different cases that are being presented and discussed. The left part of Fig. 1 displays the gas temperature profile and the normalized droplet radius for different initial droplet radii between 5 and 50 μm, and the right part shows corresponding profiles for initial droplet radii between 50 and 125 μm. For an initial droplet radius of 5 μm, the droplets enter the hot temperature region of the gas, and the small droplets quickly evaporate completely over an evaporation zone extending over about 2 mm, providing the fuel vapor for chemical reactions to take place over a wide region of about 14 mm in physical space. The spray flame is hotter than pure gas diffusion flames, which is typical for spray flames with small droplets [14]. The maximum spray flame temperature is 2,215 K, and the reaction zone just extends over the stagnation plane. A progressive increase of the initial droplet radius first leads to a decrease of the width of the reaction zone until a minimum value is reached for spray flames with an initial radius of around $R_0 = 50$ μm. As initial droplet radius is increased beyond 10 μm, two reaction zones develop at about 15 μm, where the spray sided reaction zone ends with the completion of evaporation at about -4 mm on the spray side of the configuration, and the gas flame resides on the air side of the configuration. The local minimum in the gas temperature profile resides near -4 mm, where evaporation is completed. At initial radius of 25 μm, the two peaks attain about the same flame temperature, see Fig. 2, where the maximum temperatures of the gas and spray side of the flames are plotted against the initial droplet radius. At higher initial droplet sizes, the spray crosses the stagnation plane, and it oscillates around it, see right part of Fig. 1. This phenomenon has been studied in more detail in Refs. [3, 12–14]. For an initial droplet radius between 25 and 100 μm, the spray sided flame is hotter than the gas side flame, see Fig. 2, and

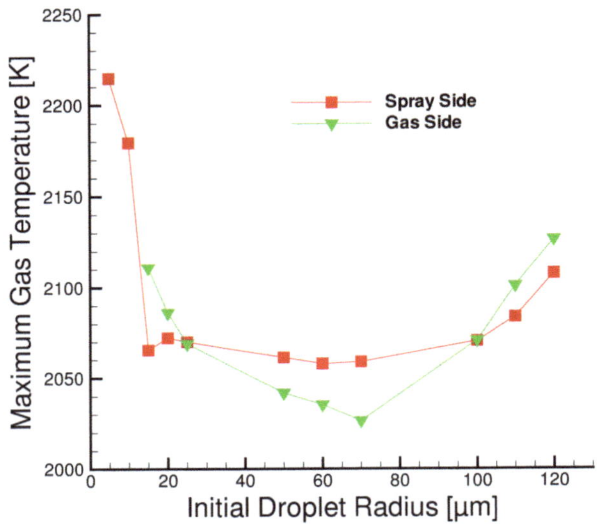

Fig. 2 Maximum gas temperature versus initial droplet radius for the spray and air sided flames

this is reversed at higher initial droplet sizes. As initial droplet size is increased, the flame moves towards the gas side of the configuration, and beyond about 125 μm, the double reaction zone disappears, and a single reaction zone is obtained, i.e. a single spray flame exists. Figure 1 also displays the normalized droplet radius, R/R_0, and it can be seen that for large droplets, they oscillate throughout the width of the reaction zone. For initial radii beyond about 125 μm, it appears that the droplet oscillation becomes so strong, that the zone of evaporation is wider than the reaction zone, and the droplets reside outside of the reaction zone on the gas side of the configuration. This leads to a flame instability in such a way that the spray evaporation cannot be achieved any more through the heat release of the chemical reactions, which eventually break down due to energy consumption of the evaporation process. The flame with the largest initial droplet size which could be obtained has an initial droplet radius of 128 μm. This extinction process is novel in the sense that the spray flame does not extinguish due to increased gas strain rate, but to increased need of energy from the gas phase for spray evaporation. The strain that leads to extinction in this situation is imposed by the droplet motion through droplet drag. This mechanism is interesting and requires more study in future, and to the authors' best knowledge, it has not been identified in the literature so far.

Figure 2 shows the maximum gas and spray sided reaction zone temperatures, and it can be seen that below an initial droplet radius of 25 μm, the gas sided flame is hotter that the spray sided flame, and between 25 and 100 μm, this is reversed. Beyond $R_0 = 100$ μm, the gas flame is hotter again due to the effects of droplet reversal and oscillation. Re-entry of the droplets enhances combustion, and therefore, the flame temperatures increase again after a local minimum is reached. The latter increase is typical for spray flames with droplet reversal [12, 13].

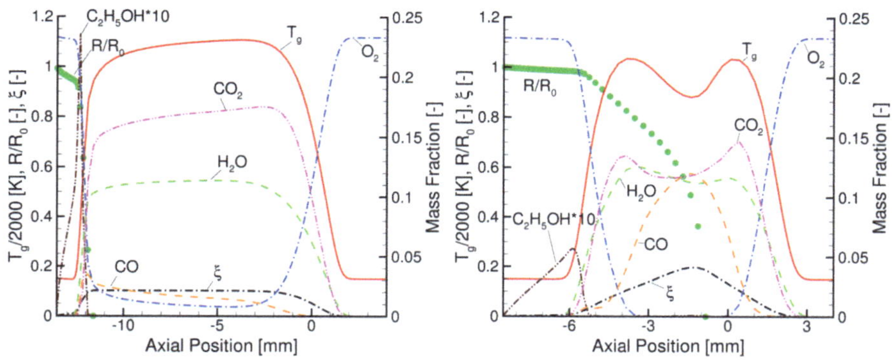

Fig. 3 Outer ethanol/air spray flame structure, a=55/s, R_0=5 μm (*left*) and R_0=25μm [23] (*right*)

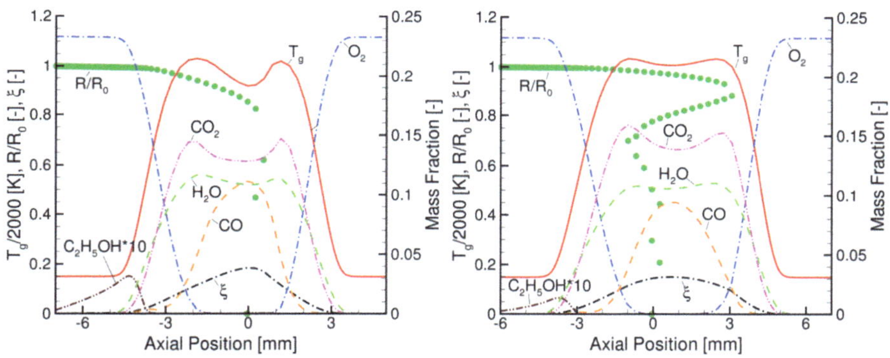

Fig. 4 Outer ethanol/air spray flame structure, a=55/s, R_0=50 μm (*left*) and R_0=100 μm (*right*)

A more detailed study of the outer and inner spray flames structures is presented in the next paragraphs. Figures 3 and 4 show the outer flame structure for four selected cases, namely for initial droplet radii of R_0=5 μm, R_0=25 μm, R_0=50 μm and R_0=100 μm. Important differences in the flame structures are found. For R_0=5 μm, evaporation occurs very fast in a very small region, and the flame tends to behave like a gas premixed flame – this is also visible from the profile of the mixture fraction, ξ, which shows an almost constant behavior over a wide range of the reaction zone. The behavior of the mixture fraction will be discussed in Sect. 3.2 in connection with the classification of different spray flame regimes. Concerning the effect of evaporation on the flame, the gas temperature profile shows a reduced flame temperature in the vaporization zone of the spray flame due to energy transfer to the spray. A progressive increase of the initial droplet radius leads to enhanced droplet penetration into the reaction zone and a displacement of the evaporation front into the main reaction is found for spray flames with initial droplet radii of R_0=25 μm and R_0=50 μm, see Figs. 3 and 4. It is observed that the width of the reaction zone decreases when the initial droplet size is increased from 5 μm to 25 μm and from 25 μm to 50 μm. However, when the initial droplet size is further increased, the width of the reaction zone

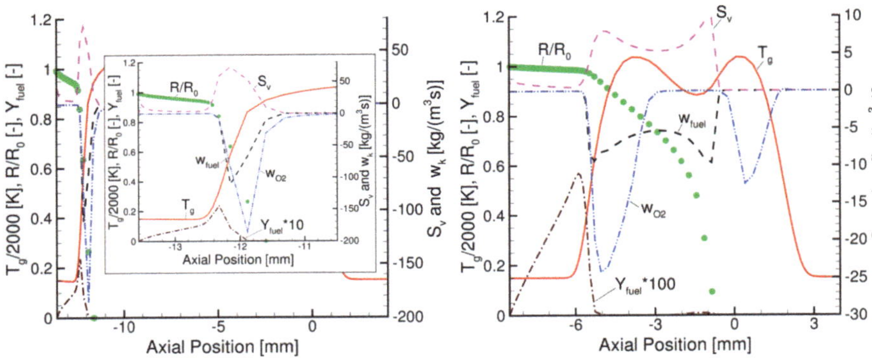

Fig. 5 Profiles of evaporation rates and chemical reaction rates, a=55/s, R_0=5 μm (*left*) and R_0=25 μm (*right*)

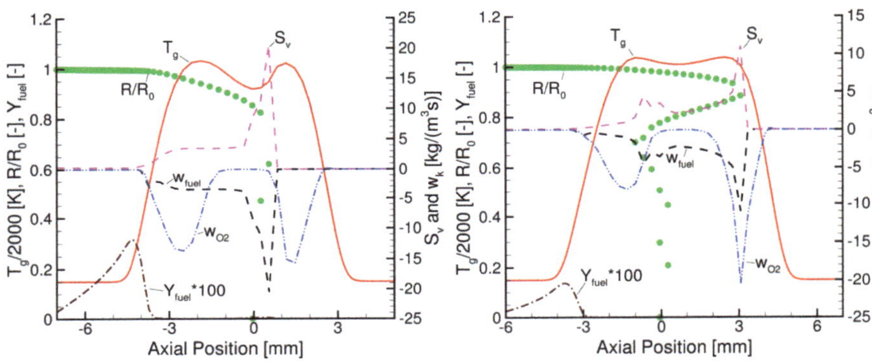

Fig. 6 Profiles of evaporation rates and chemical reaction rates, a=55/s, R_0=50 μm (*left*) and R_0=100 μm (*right*)

increases again as shown in Fig. 4 for R_0= 100 μm. This is due to droplet oscillation around the stagnation plane, which tremendously broadens the evaporation zone and therefore the flame zone. The width of the chemical reaction zone is increased following the behavior of the evaporation zone, thus demonstrating the importance of evaporation on the entire spray flame structure.

Considering the profiles of the chemical species shown in Figs. 3 and 4, typical features of gas phase chemistry such as formation of CO prior to CO_2 is observed. The principal profiles of H_2O and CO_2 follow the shape of gas temperature, whereas the profile of CO attains a maximum value in the area of local minimum of the gas temperature for high initial droplet sizes as seen in Fig. 4. The profile of the ethanol vapor is not only affected by combustion but also by evaporation as displayed in Figs. 5 and 6, where both the mass evaporation rate, S_v, and the specific reaction rates, $\dot{\omega}_k$ of oxygen and ethanol vapor are shown. Note that the left part of Fig. 5 includes a zoom of the reaction/evaporation zone on the spray side of the configuration. For R_0= 5 μm (LHS of Fig. 5), spray evaporation occurs very fast as the spray enters the reaction zone. Here, a peak of the mass evaporation rate, S_v, is found. At this location, the absolute value of the specific chemical reaction rate of the ethanol vapor is higher than the evaporation rate of liquid ethanol, indicating that molecular diffusion is important in this zone as also can be seen from the profile of the fuel vapor. Diffusion is caused by the accumulation of vapor fuel outside of the chemical reaction zone. The specific reaction rate of oxygen is also displayed, and its absolute value is about double the value of fuel vapor, which is associated with the stoichiometry of ethanol/air combustion, where about 46 kg ethanol react with about 96 kg oxygen.

For R_0= 25 μm, the spray penetrates deep into the reaction zone and two gas temperature peaks are found as well as two peaks in the profile of spray evaporation source term. As the spray enters the chemical reaction zone, a peak of the fuel vapor is found, which, however, is much smaller than the one observed for the spray with R_0= 5 μm. The fuel vapor is generated through evaporation and consumed by chemical reactions leading to relative extrema in the profiles of S_v and $\dot{\omega}_{\text{fuel}}$ at the beginning of the chemical reaction zone as well as at about -1 mm, where evaporation causes a dip in the profile of gas temperature, because the evaporation acts as an energy sink in this zone (see right part of Fig. 5). Since the chemical time is much smaller than the diffusion time in the main reaction zone, all the fuel evaporated here immediately reacts and diffusion effects of fuel vapor are very small in this zone. In the situation under consideration, the oxygen consumption occurs near the first peak of spray evaporation, whereas the second one is shifted towards the air side of the flame, and it does not coincide with the the local extremum of fuel consumption. In this region, oxygen is mainly consumed in a diffusion flame on the gas side of the counterflow configuration.

The displacement of the air sided oxygen consumption rate from the consumption rate of ethanol vapor can be explained through a more detailed study of the reaction rates for oxygen. The spray sided net consumption rate of oxygen is about -24 kg/(m^{3}s), and the major contribution of -17 kg/(m^{3}s) comes from reaction $H+O_2 \rightleftharpoons OH+O$ and the second most important contributor is the competing reaction $H+O_2+M \rightleftharpoons H_2+M$ with -12 kg/(m^{3}s). Reaction $OH+HO_2 \rightarrow O_2+H_2O$ contributes the remaining 7 kg/(m^{3}s) to the net reaction rate of oxygen in this region. The air sided peak of the oxygen net consumption rate is approximately -12.5 kg/(m^{3}s). Here, the gas temperature is almost 2100 K, which leads to the almost negligible influence of reaction $OH+HO_2 \rightarrow O_2+H_2O$. The major ethanol vapor consumption occurs in the region near the stagnation plane at -1 mm where the liquid evaporates. Here, the gas temperature is high enough for the fuel vapor to break down into radicals, which diffuse towards the air sided chemical reaction zone and there, they react with the oxygen coming from the air side of the configuration. This effect of displacement of major fuel and oxygen consumption rates is caused by spray evaporation.

For a spray with R_0= 50 μm, see left part of Fig. 6, it is observed that evaporation is delayed and the evaporation rate at the entrance of the spray into the chemical reaction zone is considerably reduced. For this reason, the vapor fuel mass fraction decreases at the peak value generated at the edge of the flame. Under these conditions, a first droplet reversal is found across the stagnation plane. At the position of droplet reversal, the residence time of the droplet is extended and a peak in the evaporation

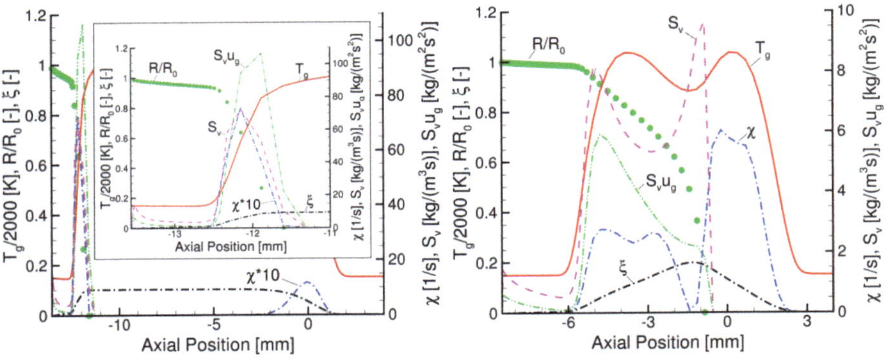

Fig. 7 Profiles of evaporation rate and scalar dissipation rates, a=55/s, R_0=5 μm (*left*) and R_0=25 μm [23] (*right*)

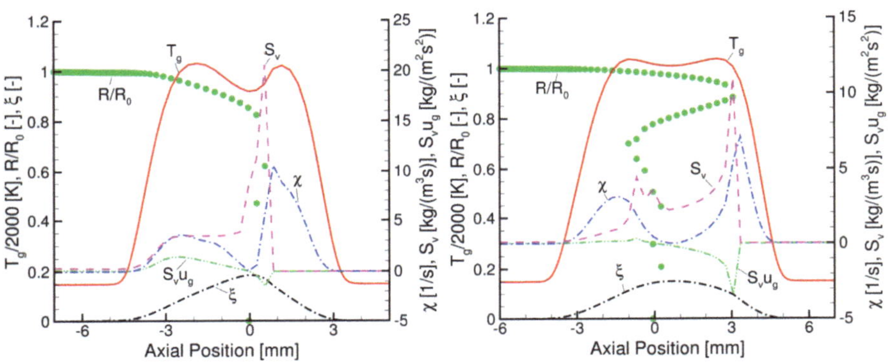

Fig. 8 Profiles of evaporation rate and scalar dissipation rates, a=55/s, R_0=50 μm (*left*) and R_0=100 μm (*right*)

rate is generated, which is much higher than the one found for R_0=25 μm. The reason for this is that the delay of droplet evaporation and the deep droplet penetration allows the droplets to evaporate in a zone of very high temperature. Figure 6 shows how evaporation is concentrated in the reversal position for R_0=50 μm, whereas its distribution is much more uniform for R_0=25 μm. For an initial droplet radius of 100 μm, the high initial momentum associated with the increased droplet mass leads to droplet oscillation. In this case, two droplet reversals are found, which are associated with two peaks in the profile of the evaporation rate. Additionally, a third peak is found at the stagnation plane. Under these conditions, the peak value of the fuel vapor concentration at the spray side of the reaction zone is very low, and it is observed that the mixture fraction profile is very uniform, see Figs. 7 and 8.

Figures 7 and 8 show the influence of the evaporation on the scalar dissipation rate profile for the four cases considered here. Note that the left part of Fig. 7 includes a zoom of the evaporation zone on the spray side of the configuration. In laminar gas diffusion flames, the profile of the scalar dissipation rate, χ, attains only one maximum value, which is located at the gas stagnation plane. In spray flames, however, the evaporation dominates the profile of the scalar dissipation rate as discussed for methanol/air spray flames [17]. For small droplet sizes, the droplets do not reach the stagnation plane, and the position of the peak value of the scalar dissipation rate is maintained at the stagnation plane, since evaporation does not take effect [13]. However, in the zone where evaporation takes place, the profile of the scalar dissipation rate strongly differs from the one that is characteristic for gas flames. In general, the evaporation mass source generates different local mixture states leading to considerably different values of the gradient of the mixture fraction, where high local values of the scalar dissipation rate are generated, see Eq. (23). For R_0= 5 μm, a peak in the profile of the scalar dissipation rate, additional to the one located at the stagnation plane, is found in the spray zone, whereas for R_0= 25 μm, two extra peaks are found in this area. When the droplet penetration is increased and the droplets cross the stagnation plane, the peak located at the stagnation plane is displaced towards the gas side of the configuration, and it coincides with the location of the droplet reversal position [17]. For R_0= 100 μm, two droplet reversals occur, which are associated with the position of the maxima in the profile of the scalar dissipation rate. The importance of the maximum local values of the evaporation rate generated at the positions of droplet reversal and the associated local maximum of the scalar dissipation rate will be discussed in the next section. Figures 7 and 8 show that the local maximum values of the scalar dissipation rate decrease when the initial droplet radius is increased. This is related to the more uniform profile of mixture fraction generated by large initial droplets. However, as shown in Figs. 3 and 4, an increase of the initial droplet radii also reduces the fuel vapor concentration due to the lower residence time.

A further increase of the initial droplet radius to 128 μm leads to the extinction of the flame because the evaporation zone on the air side of the flame moves towards the air inlet and leaves the reaction zone where the gas temperature becomes too low to sustain droplet evaporation, see the right part of Fig. 1. Thus, the mechanism of spray flame extinction is caused by a breakdown of the evaporation zone due to lack of energy transfer from the gas to the spray, which is caused by the fact that the reaction and the evaporation zones do not coincide any longer. The physical process of spray flame extinction is quite different from the flame extinction process caused by increased strain, where both evaporation and combustion zones coincide and chemical reactions break down because of the reduced residence time of chemical species.

3.1.2 Evaporation Effects at Extinction Strain Rates

In this section, laminar spray flame structures in the physical space are generated for different strain rates, from 55/s up to extinction. Four different initial droplet radii are considered, namely R_0= 5 μm, R_0= 25 μm, R_0= 50 μm and R_0= 100 μm.

Figure 9 shows the maximum flame temperature versus gas strain rate on the spray side of the counterflow configuration for the different initial droplet sizes under

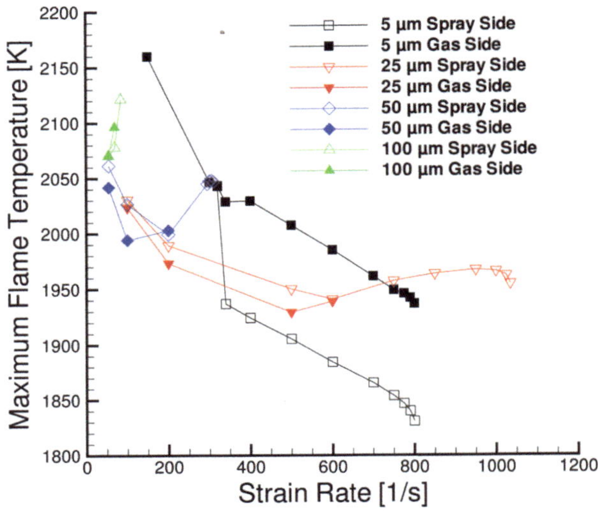

Fig. 9 Maximum gas temperature versus gas strain rate on the spray side for different initial droplet radii

consideration. Open symbols show maximum flame temperatures on the spray side and filled symbols on the air side of the configuration, and the lines are drawn for a better visibility of similar conditions. The hottest flame occurs for the smallest initial droplet radius of 5 μm. For low strain, one single reaction zone exists, see Fig. 3, and for strain rates higher than 300/s, a double flame develops where the spray sided flame is always considerably colder than the gas flame on the gas side of the configuration, see Fig. 10 left part. At a strain rate of 800/s, the reactions break down on the spray side of the configuration, and extinction occurs for both reaction zones. For an initial droplet radius of 25 μm, two reaction zones exist at low strain, c.f. Fig. 3 right part, where the gas sided flame is somewhat colder than the flame on the spray side, which is a consequence of the deep penetration of the spray into the reaction zone. The spray flames become narrower with increased strain, and at 600/s, the reaction zones merge to yield a single one as shown in Fig. 9. Flame temperature increases between a strain rate of 500/s and about 1,000/s due to droplet oscillation, and beyond 1,000/s it decreases because of reduced residence time. Extinction of the single reaction zone occurs at 1,035/s. Figure 9 shows that the spray flames with an initial droplet radius of 25 μm are the most stable compared to the other conditions studied. The spray flames with an initial droplet radius of 50 μm behave similarly to the 25 μm situation except that the extinction strain rate is much lower, namely 310/s. Figure 9 shows that the maximum flame temperature for R_0= 100 μm increases with higher strain rate until extinction is suddenly reached at a strain rate of 85/s. The gas sided flame is somewhat hotter than the spray sided flame, which is typical for sprays with a large initial droplet radius penetrating deeply into the gas sided chemical reaction zone. The increase of flame temperature with higher initial droplet size has also been observed by Gutheil and Sirignano [14]. In summary, it

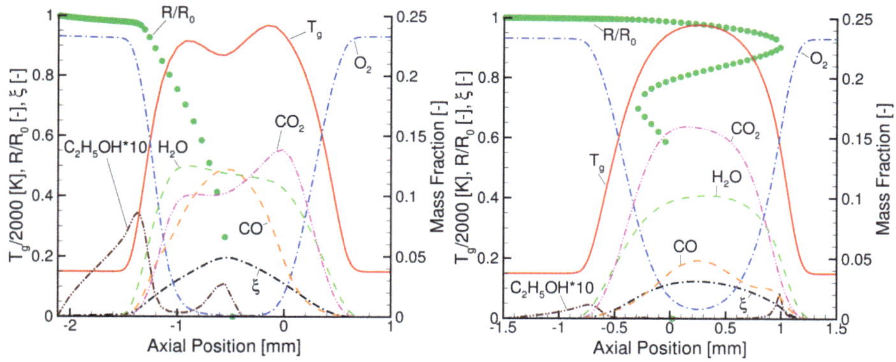

Fig. 10 Outer ethanol/air spray flame structure, a= 800/s, R_0=5 μm (*left*) and a= 1.035/s, R_0= 25 μm (*right*)

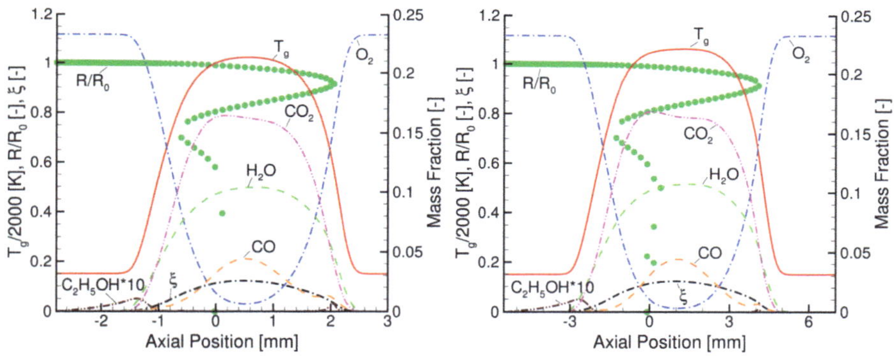

Fig. 11 Outer ethanol/air spray flame structure, a= 310/s, R_0=50 μm (*left*) and a= 85/s, R_0= 100 μm (*right*)

can be seen that the structure of spray flames in the counterflow configuration and their extinction behavior is greatly dominated by the spray process. In particular, the penetration depth of the spray into the spray sided and the gas sided reactions zones plays a major critical role: the spray penetration depth and possible oscillation determine if the spray or the gas sided flames are hotter and if there are one or two reaction zones in a flame [11, 14]. The study shows that for a small initial droplet size, a single reaction zone may be found which develops into two reaction zones as strain rate is increased, whereas the opposite behavior occurs for intermediate droplet sizes, because the spray flame narrows and the reaction zones merge. Spray oscillation favors two different reaction zones. This behavior, however, would still allow for multiple solutions of spray flames for low strain rates as discussed in Ref. [13].

The outer spray flame structure at extinction for the four different initial droplet radii discussed in this section is shown in Figs. 10 and 11. Figures 12 and 13 show the corresponding profiles of the evaporation rate and chemical reaction rates, and Figs. 14 and 15 display the evaporation rate and the scalar dissipation rate. The spray

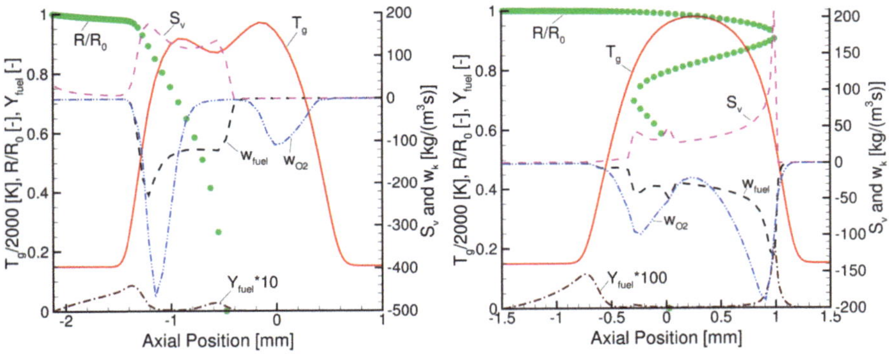

Fig. 12 Profiles of evaporation rate and chemical reaction rates, a= 800/s, R_0= 5 μm (*left*) and a= 1035/s, R_0= 25 μm (*right*)

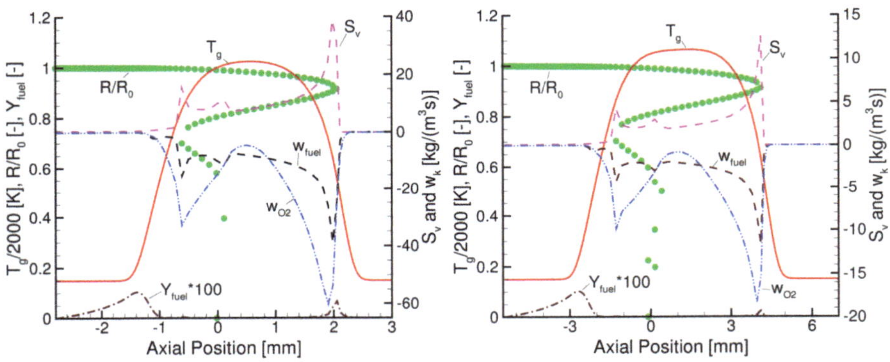

Fig. 13 Profiles of evaporation rate and chemical reaction rates, a= 310/s, R_0= 50 μm (*left*) and a= 85/s, R_0= 100 μm (*right*)

flame structures are shown for the strain rate at extinction, see Fig. 9. For R_0= 5 μm, a local minimum value of the temperature profile is located in the center of the reaction zone (see left part Fig. 12), which is similar to the situation for intermediate droplet radii at low strain, c.f. Fig. 3. However, in the low strain result for larger initial droplet size, all fuel vapor is consumed in this region in contrast to the present high strain rate result, where a considerable amount of fuel is present due to reduced residence time of the reactants. Chemical reactions are retarded, which is also visible in the profile of CO_2, where the first local maximum is much smaller than the second one close to the air side of the configuration. This retardation is also reflected in the profile of CO, which attains a local maximum where the dip in the gas temperature profile occurs, and here, CO reaches higher values than CO_2. As initial droplet size is increased, spray oscillation occurs and broadens the spray flame, and a single reaction zone is visible, which moves towards the gas side of the configuration.

Figure 12 shows the profiles of the chemical reaction rate of the fuel vapor and the evaporation rate, revealing that evaporation commences prior to chemical reaction,

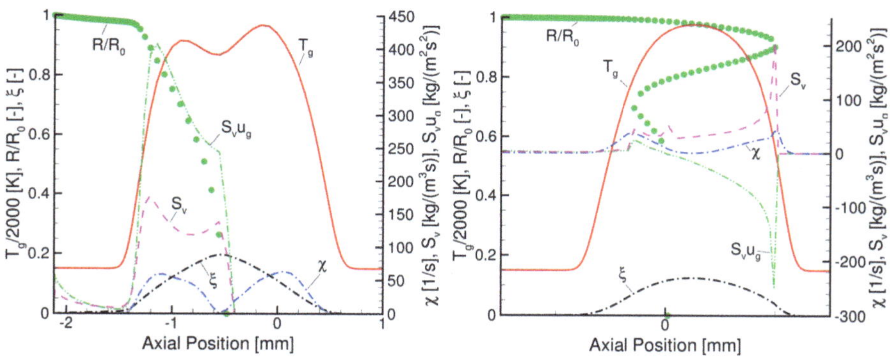

Fig. 14 Profiles of evaporation rate and scalar dissipation rates, $a = 800/\text{s}$, $R_0 = 5\ \mu\text{m}$ (*left*) and $a = 1{,}035/\text{s}$, $R_0 = 25\ \mu\text{m}$ (*right*)

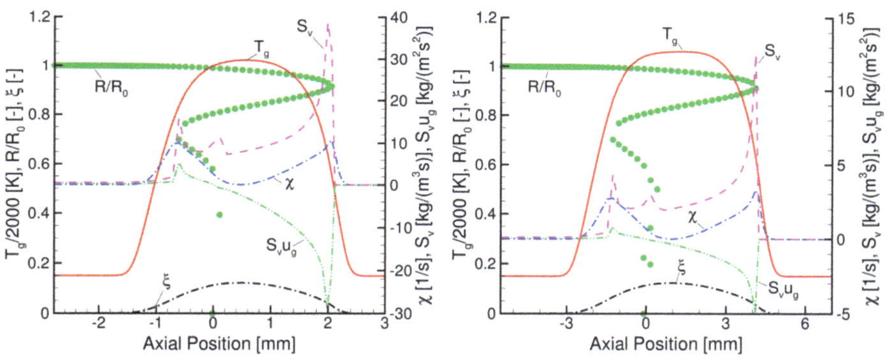

Fig. 15 Profiles of evaporation rate and scalar dissipation rate, $a = 310/\text{s}$, $R_0 = 50\ \mu\text{m}$ (*left*) and $a = 85/\text{s}$, $R_0 = 100\ \mu\text{m}$ (*right*)

which, initially overweighs evaporation before it balances the evaporation rate. This phenomenon is produced by the peak of ethanol vapor observed before the chemical reaction zone. The initial peak of the vapor mass fraction produces important fuel vapor diffusion from the edge of the flame to the main reaction zone. It is observed that the width of the reaction zone is increased when the initial droplet size is increased, which is associated with the increase in the oscillation of the spray. This phenomenon is attributable to the higher momentum associated with the increased droplet size which enhances droplet penetration. The flame structure for $R_0 = 25\ \mu\text{m}$ is completely different compared to the case of $R_0 = 5\ \mu\text{m}$. Droplet oscillation is found, generating two droplet reversals at the edges of the reaction zone. Under these conditions, the fuel vapor mass peak located at the entrance of the reaction zone is reduced and the second peak observed in the dip of the temperature profile for 5 μm, is shifted towards the gas side of the configuration to the position of the droplet reversal. When the initial droplet size is further increased, the fuel concentration decreases due to the reduced residence time (see Fig. 11), and extinction is reached. The profile of the chemical reaction rate of oxygen attains two local minima as in the low strain results, and they reside in the gas and the spray sides of the configuration where chemical reactions are initiated.

Figures 14 and 15 show the scalar dissipation rate profile for the four different initial droplet radii studied. For $R_0 = 5\ \mu$m, two local maxima are found in the profile of the scalar dissipation rate, which reside on each side of the configuration. Under the conditions investigated, the droplets do not reach the stagnation plane, and therefore, the gas sided peak of the scalar dissipation rate is found at the gas stagnation plane, which is typical for gas phase combustion. Figures 14 and 15 show a progressive reduction of the local maximum values of the evaporation rate under extinction conditions for an increase of the initial droplet size, and their local maxima coincide with the left and right boundary of the evaporation zone if droplet reversal occurs. In these regimes, the evaporation is strongest as can be seen from the profile of the evaporation rate, a third smaller peak in the evaporation rate is found at the stagnation plane around which the droplets oscillate. At the positions of droplet reversal, the residence time is longest, and therefore, evaporation is enhanced. The same is true for the gas stagnation plane with the droplets oscillating around it. Evaporated fuel vapor is transported away from the location of evaporation through gas phase motion, and the local maxima of the profile of the product of the gas velocity, u_g, and the evaporation rate, S_v, coincide with the local maxima of the scalar dissipation rate, χ, which no longer resides at the stagnation plane as in gaseous counterflow diffusion flames. The reduced values of the scalar dissipation rate with increased initial droplet size are related to the increase in gas temperature, which is comparable to the fact that scalar dissipation rate increases with increased strain rates in gas flames, which is accompanied by a reduction of the flame temperature.

3.2 Multi-Regime Spray Combustion

The differences in spray flame structure discussed so far induce different spray combustion regimes, which are discussed in this section. For this purpose, flame structures at low strain rate are studied for initial droplet radii of 5, 10, 15 and 20 μm. Figures 16 and 17 show the gas temperature, the mixture fraction and the normalized droplet radius in physical space. Considering the results for an initial droplet radius of 5 μm, the profile of the mixture fraction motivates the characterization of three different regimes of combustion. A first zone, where the mixture fraction increases monotonically with space, is generated by spray evaporation. Once the droplets are fully evaporated, the mixture fraction attains an almost constant value before it decreases due to extended mixing with the opposed air stream. The major reaction zone, where the mixture fraction attains an almost constant value, may be considered premixed-like, and in the third regime, where considerable mixing with the counter flowing air occurs, is non-premixed. The present flamelet equations derived in Sect. 2.2 are valid in the first and third zones, where the contribution of spray evaporation is present in the first regime of the flame shown in the left part of Fig. 16. The intermediate

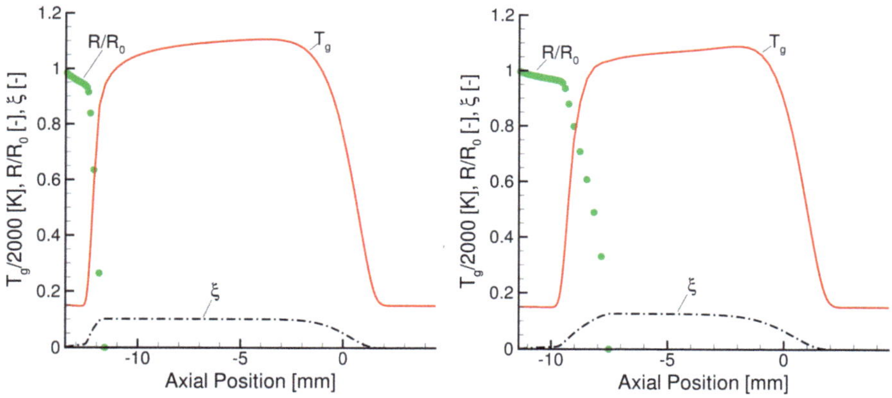

Fig. 16 Mixture fraction, gas temperature, and normalized droplet radius for an ethanol/air spray flame in physical space, $a = 55$/s, $R_0 = 5$ μm (*left*) and $R_0 = 10$ μm (*right*)

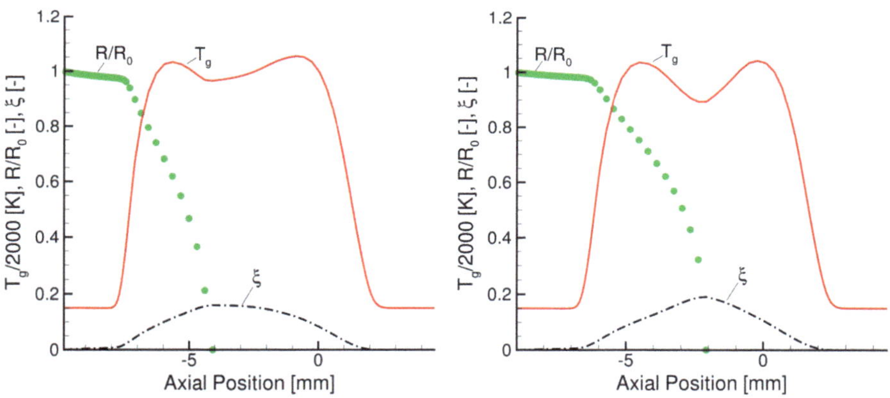

Fig. 17 Mixture fraction, gas temperature, and normalized droplet radius for an ethanol/air spray flame in physical space, $a = 55$/s, $R_0 = 15$ μm (*left*) and $R_0 = 20$ μm (*right*)

range with considerable premixed combustion is not directly affected by evaporation [13], and here, pure gas flamelet formulations such as the ones discussed in the introduction, may be applied [7, 10, 18, 22, 24–26].

The major improvement with respect to the present new spray flamelet formulation would act in the first part of the discussed regime, where the spray evaporates quickly and the spray does not penetrate deep into the combustion zone. When droplet penetration is enhanced as it is the case for larger initial droplet sizes and higher strain rates, the first non-premixed-like zone is extended and the premixed-like zone is reduced (see Figs. 16 right part and 17). Spray penetration increases both with initial droplet size and droplet velocity and with gas strain rate, which is typical for spray combustion. At higher strain rates and for larger initial droplet size, the droplets penetrate the stagnation plane and they may oscillate around it as discussed in Sects. 3.1 and 3.1.1, leading to an increased importance of the present new spray flamelet formulation. This will be discussed in Sect. 3.3.

Figures 16 and 17 show a gradual increase in initial droplet size with a fixed equivalence ratio and inlet velocity, and it can be seen that the second regime identified above as premixed-like combustion, becomes less relevant and eventually disappears, so that two non-premixed flame regimes reside, where the spray sided regime accommodates a reaction zone directly interacting with the spray and the gas side of the configuration embeds a gas flame. For mono-disperse sprays with an initial droplet radius of 20 μm, the premixed-like region completely disappears, and two diffusion flames exist where the spray-sided flame is a spray diffusion flame and the gas-sided reaction zone is a pure gas diffusion flame. This situation is also found as strain rate is increased, see Sect. 3.1.1, and the situation, where both flames merge, is also a non-premixed situation. As strain rate is increased, flame extinction may occur, which is used for the implementation of (spray) flamelet modeling [13, 17, 24].

It can be concluded that the present flamelet formulation is appropriate for regimes where non-premixed combustion dominates, which is the case for spray flames beyond about 10–15 μm initial droplet radii at low strain and smaller initial radii at elevated strain rates. The importance of spray evaporation in the spray-controlled zone is evident and therefore, a proper evaluation of the capability of the present revised flamelet formulation for non-premixed spray flames is required.

3.3 Evaporation Effects in Mixture Fraction Space

This section provides spray flame structures at both low and high strain rate in mixture fraction space, and the relevance of the new spray flamelet formulation with respect to the contribution of evaporation source terms is evaluated. For this purpose, two flame structures for a mono-disperse spray flame of 25 μm initial droplet radius are chosen at low (55/s) and high (950/s) strain rates. A droplet radius of 25 μm is selected in order to ensure a spray flame with no premixed-like zones so that the present model is applicable in all flame zones.

Numerical simulations of counterflowing ethanol/air spray flames are performed using a similarity transformation as described by Continillo and Sirignano [5] and extended by Gutheil and Sirignano [14] as well as by Gutheil [13]. A very fine grid of 1,000 nodes is used for the present simulations. The right parts of Figs. 3 and 7 show the low strain result in physical space, and Fig. 18 displays the flame structure at 950/s, which is prior to extinction occurring at 1,035/s, c.f. Sect. 3.1.1. In both cases, the spray flames do not include premixed regimes. The spray penetrates into the reaction zone, and for elevated strain rate, droplet oscillation around the stagnation plane occurs. In the latter situation, the spray is no longer mono-disperse as seen in Fig. 18. Two reaction zones prevail for the low strain situation, where the local minimum is caused by spray evaporation, and the reaction zones merge in the case of the higher strain rate [13]. The profiles of the evaporation rate and the scalar dissipation rate are displayed in the right part of Fig. 18 for the high strain rate,

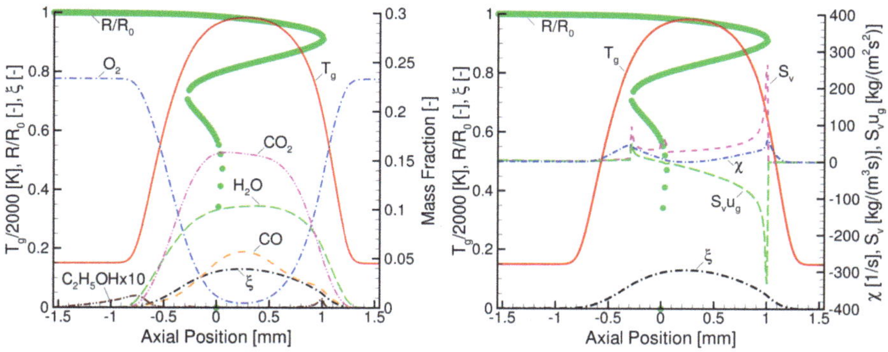

Fig. 18 Outer ethanol/air spray flame structure, $R_0 = 25\,\mu\text{m}$, a = 950/s [23]

whereas corresponding low strain results are displayed in the right part of Fig. 7. In order to study the influence of spray evaporation on the flamelet formulation in Eq. (24), the flame structures are presented and evaluated for different species in the mixture fraction space.

Figure 19 shows the contribution of the dissipation, mixing/evaporation and evaporation terms to the flamelet Eq. (24) of the mass fraction of ethanol vapor for the low strain situation, where the left side shows the spray side and the right side the gas side of the configuration. This distinction is necessary because of the non-monotonicity of mixture fraction with space. For the present analysis, the local maximum of the mixture fraction is used to separate the figures. The results shown for the spray side of the configuration therefore show an increasing scale of the mixture fraction whereas the gas-sided structures show a decrease of mixture fraction on the abscissa. On the spray side of the configuration, the dominating term defined in Eq. (24) is found to be the pure evaporation term. The dissipation term is relevant at the beginning of the evaporation zone, where chemical reactions are initiated at low values of the mixture fraction. Here, the slope of the profile of the ethanol vapor mass fraction changes due to evaporation (leading to an increase in fuel vapor mass fraction) and starting chemical reactions (decreasing fuel vapor mass fraction), see Fig. 3. At higher values of mixture fraction, the evaporation term dominates the flamelet equation. In this region, the mixing/evaporation term plays a minor role.

The right part of Fig. 19 displays the same profiles for the gas side of the configuration. The droplets are completely evaporated at high values of the mixture fraction, which leads to a dominance of the evaporation term just where evaporation is completed. Since there is hardly any evaporated fuel left, all contributions are about zero on the gas side of the configuration.

Corresponding profiles at high strain rates are shown in Fig. 20. Under these conditions, the evaporation is dominant again, and the dissipation term becomes somewhat relevant in the first position of droplet reversal, but the pure evaporation term strongly dominates the structure. The gas side evaluation plotted in Fig. 20, right part, shows that the presence of droplets changes the findings for the low strain

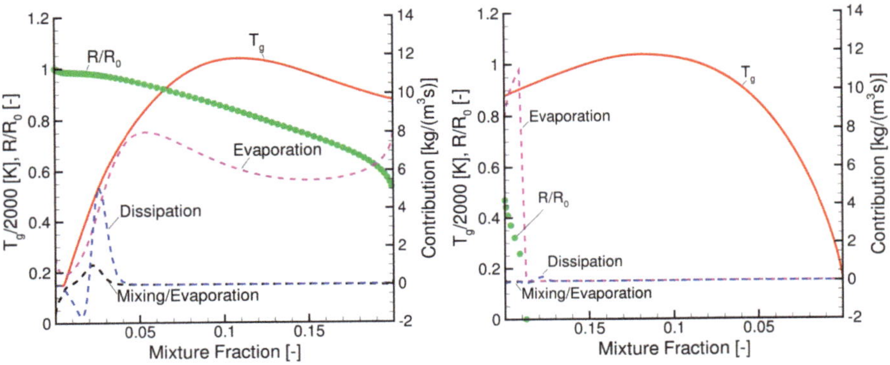

Fig. 19 Contributions for ethanol, spray side (*left*) and gas side (*right*), a=55/s, R_0=25 μm [23]

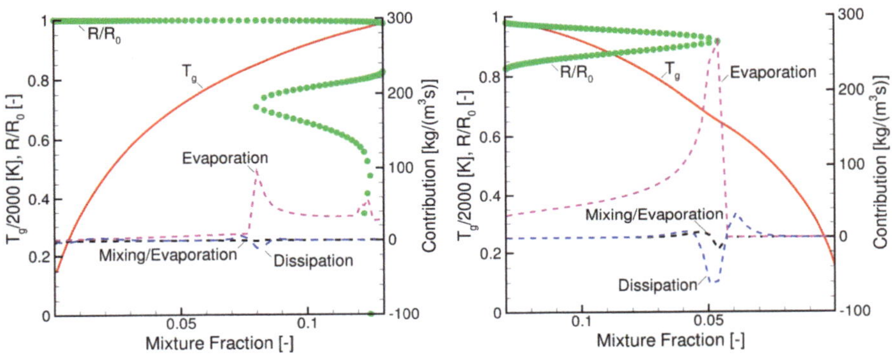

Fig. 20 Contributions for ethanol, spray side (*left*) and gas side (*right*), a=950/s, R_0=25 μm [23]

situation in such a way that the evaporation again shows significant influence on the flamelet equation of the mass fraction of fuel vapor, whereas the pure dissipation term plays a minor role. Overall, it can be stated that the flamelet equation for the fuel vapor is dominated by the evaporation term of the flamelet equation, and this explains why approaches neglecting this term are not suitable to represent the flamelet structure of laminar spray flames with gas flamelet models, and evaporation must be taken into account in spray flamelet computations of turbulent spray flames.

Figure 21 shows the low strain results for the flamelet equation of O_2 and the corresponding results for high strain conditions are displayed in Fig. 22. In both situations, the mixing/evaporation term dominates the flamelet equation on the spray side of the configuration, and the dissipation term also shows a considerable contribution. The evaporation term alone is not relevant, because oxygen is not an evaporating component. Even though this is the case, it can be seen that the effect of evaporation on the spray flamelet equation for the mass fraction of oxygen may not be neglected, and it has a pronounced influence through the combined mixing/evaporation term. For the low strain situation (c.f. Fig. 21), the droplets do not cross the stagnation

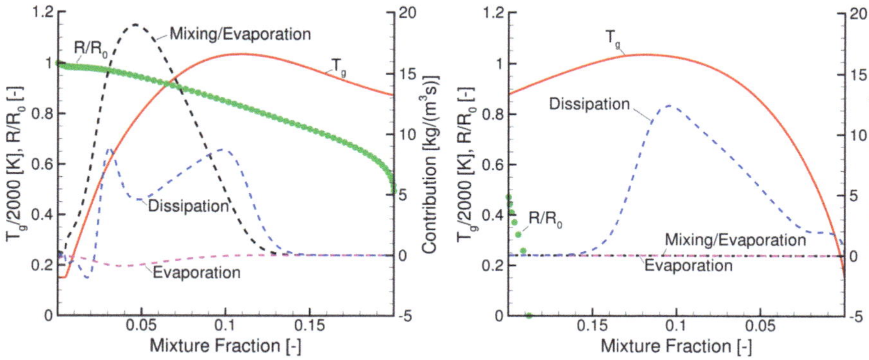

Fig. 21 Contributions for O_2, spray side (*left*) and gas side (*right*), a=55/s, R_0=25 μm [23]

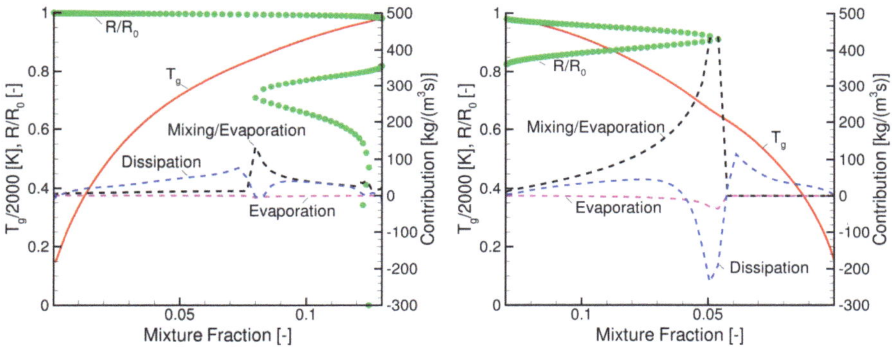

Fig. 22 Contributions for O_2, spray side (*left*) and gas side (*right*), a=950/s, R_0=25 μm [23]

plane, and dissipation determines the flamelet equation for oxygen on the gas side of the configuration, note that the pure gas combustion situation is recovered. For increased strain, the droplets cross the stagnation plane, and the mixing/evaporation term again dominates the equation with considerable contribution also of dissipation.

Corresponding profiles for the flamelet equation of CO_2, are shown in Figs. 23 and 24. Since CO_2 is produced whereas O_2 is consumed and both are non-evaporating components, the contributions of the different terms are similar, however, their sign is opposite.

In summary, it can be concluded that for the evaporating fuel ethanol, the evaporation term in the flamelet equation is dominating wherever the spray is present. The flamelet equations of oxygen and carbon dioxide are dominated by the contribution of combined mixing/evaporation, whereas the contributions have opposite signs since oxygen is consumed and carbon dioxide is produced. At a low strain rate where the droplets do not cross the stagnation plane, the gas flamelet contribution is recovered on the gas side of the configuration.

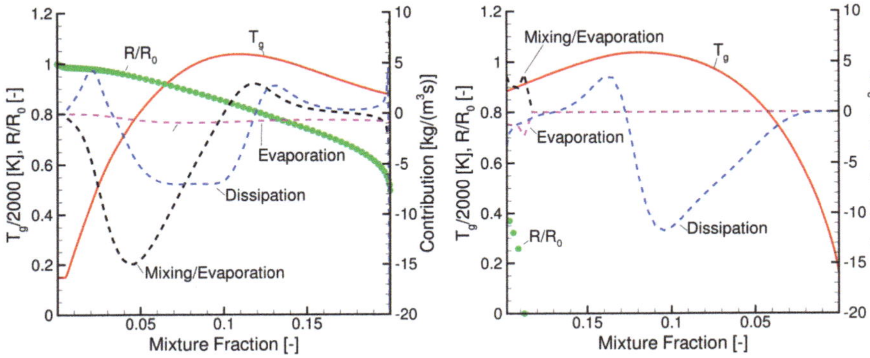

Fig. 23 Contributions for CO_2, spray side (*left*) and gas side (*right*), a=55/s, R_0=25 μm [23]

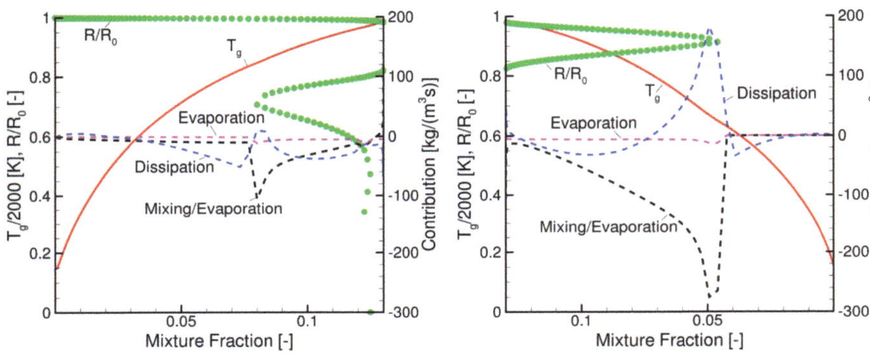

Fig. 24 Contributions for CO_2, spray side (*left*) and gas side (*right*), a=950/s, R_0=25 μm [23]

The present results show that the terms attributable to spray processes are important under all conditions, and they must be always considered. The spray flamelet equations derived and evaluated in the present study are suitable to describe spray diffusion flames in the counterflow configuration. Spray evaporation is taken into account and dominating evaporation effects are found for the evaporating component, whereas mixing/evaporation effects are relevant for non-evaporating components with contributions of the dissipation term. If droplet oscillation occurs, the gas side of the configuration is also dominated by the latter terms. In situations without droplet oscillation, the flamelet formulation for pure gas combustion is recovered. Thus, the present formulation is an appropriate extension of the classical flamelet formulation for gas flames to spray flames.

4 Conclusions

Laminar mono-disperse ethanol/air spray flame structures in an axi-symmetric counterflow configuration have been studied by means of numerical simulations. The multi-regime nature of spray flames has been illustrated, and it is found that in most situations, the combustion regime is dominated by evaporation or combined mixing/evaporation, which cannot be characterized by gas flamelet formulations.

A new set of spray flamelet equations has been derived in order to properly account for the evaporation process in these flames. They recover the gas flamelet equations as a special case. The relative importance of terms in the spray flamelet equation attributable to evaporation has been evaluated for different species including the evaporating fuel ethanol, oxygen, and the product carbon dioxide. The results show that the effects neglected by gaseous flamelet equations are important in all cases studied, and therefore they must be included in revised multi-regime flamelet methods. The present results suggest a revised flamelet model for use in turbulent spray flame computations using the mixture fraction, its scalar dissipation rate and the spray mass evaporation term as dependent variables. This formulation includes the need for the determination of a joint probability density function depending on these variables, which will be the scope of future research.

Acknowledgement H.O. thanks the DAAD for financial support through a research fellowship. Financial support of the German Research Foundation through the Graduate School "MathComp" of IWR is gratefully acknowledged.

References

1. Abramzon, B. and Sirignano, W. A., *Droplet vaporization model for spray combustion calculation*, Int. J. Heat Mass Transfer 32(9) (1989), pp. 1605–1618.
2. Chen, Z. H., Lin, T. H. and Sohrab, S. H., *Combustion of Liquid Fuel Sprays in Stagnation-Point Flow*, Combust. Sci. Tech. 60 (1988), pp. 63–77.
3. Chen, N. H., Rogg, B., and Bray, K. N. C., *Modelling laminar two-phase counterflow flames with detailed chemistry and transport*, Proc. Combust. Inst. 24 (1992), pp. 1513–1521.
4. Chevalier, C., *Entwicklung eines detallierten Reaktionsmechanismus zur Modellierung der Verbrennungsprozesse von Kohlenwasserstoffen bei Hoch- und Niedertemperaturbedingungen*, University of Stuttgart, Stuttgart (1993).
5. Continillo, G. and Sirignano, W. A., *Counterflow Spray Combustion Modeling*, Combust. Flame 81 (1990), pp. 325–340.
6. Dhuchakallaya, I., Rattanadecho, P., Watkins, P., *Auto-ignition and combustion of diesel spray using unsteady flamelet model*, Applied Thermal Engineering 52 (2013), pp. 420–427.
7. Franzelli, B., Fiorina, B., Darabiha, N., *A tabulated chemistry method for spray combustion*, Proc. Combust. Inst. 34 (2013), pp. 1659–1666.
8. Ge, H. W. and Gutheil, E. , *Simulation of a Turbulent Spray Flame using coupled PDF Gas Phase and Spray Flamelet Modeling*, Combust. Flame 153 (2008), pp. 173–185.
9. Ge, H. W., Düwel, I., Kronemayer, H., Dibble, R. W., Gutheil, E., Schulz, C., Wolfrum, J., *Laser-Based Experimental and Monte Carlo PDF Numerical Investigation of an Ethanol/Air Spray Flame*, Combust. Sci. Tech. 135 (2008), pp. 1529–1547.
10. Gicquel, O., Darabiha, N., Thévenin, D., *Laminar premixed hydrogen/air counterflow flame simulations using flame prolongation of ILDM with differential diffusion*, Proc. Combust. Inst. 28 (2000), pp. 1901–1908.
11. Greenberg, J. B., and Sarig, N., *An Analysis of Multiple Flames in Counterflow Spray Combustion*, Combust. Flame 104 (1996), pp. 431–459.
12. Gutheil, E., *Structure and extinction of laminar ethanol-air spray flames*, Combust. Theory and Modelling 5(2) (2001), pp. 131–145.

13. Gutheil, E., *Multiple Solutions for Structures of Laminar Counterflow Spray Flames*, Prog. Comput. Fluid Dyn. 5(7) (2005), pp. 414–419.
14. Gutheil, E. and Sirignano, W. A., *Counterflow Spray Combustion Modeling with Detailed Transport and Detailed Chemistry*, Combust. Flame 113 (1998), pp. 92–105.
15. Gutheil, E., Williams, F. A., *A Numerical and Asymptotic Investigation of Structures of Hydrogen-Air Diffusion Flames at Pressures and Temperatures of High-Speed Combustion*, Prog. Combust. Inst. 23 (1990), pp. 513–521.
16. Hasse, C. and Peters, N., *A two mixture fraction flamelet model applied to split injections in a DI Diesel engine*, Proc. Combust. Inst. 30 (2005), pp. 2755–2762.
17. Hollmann, C. and Gutheil, E., *Flamelet-Modeling of Turbulent Spray Diffusion Flames Based on a Laminar Spray Flame Library*, Combust. Sci. Tech. 135 (1998), pp. 175–192.
18. Knudsen, E., Pitsch, H., *A general flamelet transformation useful for distinguishing between premixed and non-premixed modes of combustion*, Combust. Flame 156 (2009), pp. 678–696.
19. Li, S. C., Libby, P. A. and Williams, F. A. *Experimental and theoretical studies of counterflow spray diffusion flames*, Proc. Combust. Inst. 24 (1992), pp. 1503–1512.
20. Li, S. C., *Spray Stagnation Flames*, Prog. Energy Combust. Sci. 23 (1997), pp. 303–347.
21. Massot, M., Kumar, M., Smooke, M. D. and Gomez, A., *Spray counterflow diffusion flames of heptane: Experiments and computations with detailed kinetics and transport*, Proc. Combust. Inst. 27 (1998), pp. 1975–1983.
22. Nguyen, P., Vervish, L., Subramanian, V. Domingo, P. *Multidimensional flamelet-generated manifolds for partially premixed combustion*, Combust. Flame 127 (2010), pp. 43–61.
23. Olguin, H., Gutheil, E., *Influence of evaporation on spray flamelet structures*, Combust. Flame (2013), http://dx.doi.org/10.1016/j.combustflame.2013.10.010.
24. Peters, N., *Laminar diffusion flamelet models in non-premixed turbulent combustion*, Prog. Energy Combust. Sci. 10 (1984), pp. 319–339.
25. Pitsch, H., Peters, N., *A consistent flamelet formulation for non-premixed combustion considering differential diffusion effects*, Combust. Flame 114 (1998), pp. 26–40.
26. van Oijen, J. A., Lammers, F. A., de Goey, L. P. H., *Modeling of complex premixed burner systems by using flamelet-generated manifolds*, Combust. Flame 127 (2001), pp. 2124–2134.

Large Eddy Simulation of Diluted Turbulent Spray Combustion Based on FGM Methodology: Effect of Fuel and Mass Loading

Amsini Sadiki, Mouldi Chrigui, Fernando Sacomano and Assaad R. Masri

Abstract A numerical methodology relying on Large Eddy Simulation is used to analyze and evaluate the impact of fuel and mass loading on turbulent spray combustion. To retrieve the flow, mixing and combustion proper-ties, an Eulerian-Lagrangian approach is adopted. The method includes a full two-way coupling between the interacting two phases in presence, while the evaporation process is described by a non-eqnilibrium vaporization model. The carrier phase turbulence is captured by a combustion LES technique in which first order sub-grid scale models are applied.

Two different fuels are used to produce spray jets through a pilot flame and a co-flowing atmospheric air. A spray pre-evaporation zone enables the combustion regime to turn from diffusion to partially premixed mode. The first liquid fuel is acetone, preferred for its ability to vaporize quickly. It is modeled by a detailed reaction mechanism including 84 species and 409 elementary reactions. The ethanol as second fuel is widely used as alternative fuel. It is modeled by a detailed reaction mechanism consisting of 56 species and 351 reversible reactions. To reduce the computational costs, the combustion is described by means of a detailed tabulated chemistry approach according to the Flamelet Generated Manifold (FGM) strategy. The occurring flow and combustion properties are numerically analyzed and compared with experimental data for both fuels under different mass loading conditions. The impact of fuel and mass loading on turbulent spray combustion is evaluated in terms of flame structure, exhaust gas temperature, droplet velocities and diameters, droplet velocity fluctuations, and spray volume flux at different distances from the exit planes.

A. Sadiki (✉) · M. Chrigui · F. Sacomano
Institute for Energy and Power plant Technology, Department of Mechanical Engineering, TU Darmstadt, Jovanka-Bontschits-Str. 2, 64287 Darmstadt, Germany
e-mail: sadiki@ekt.tu-Darmstadt.de

M. Chrigui
Research Unit Materials, Energy and Renewable Energies, University of Gafsa, Gafsa, Tunisia

A. R. Masri
School of Aerospace, Mechanical and Mechatronic Engineering, The University of Sydney, Sydney, NSW 2006, Australia

B. Merci, E. Gutheil (eds.), *Experiments and Numerical Simulations of Turbulent Combustion of Diluted Sprays,* ERCOFTAC Series 19, DOI 10.1007/978-3-319-04678-5_5,
© Springer International Publishing Switzerland 2014

1 Introduction

Advanced low emission combustion chamber strategies are under development in order to meet the necessity for transportation and power generation industries to fulfill stringent regulations concerning pollutants emissions. These concepts are sensitive to variable time- and space uniformity of fuel vapor composition inherent to liquid fuels used. These time- and space varying fuel properties (in the vapor and in liquid phase) affect substantially the vaporization and kinetics-related processes, like ignition, flame propagation/stability and pollutants level. As such issues when designing combustion systems for liquid fuel are essential for the understanding of flame ignition and extinction and the prediction of pollutant formation and emission [1, 3, 14, 17, 19, 24, 27, 28, 30, 31, 40, 43, 45, 49, 52], an accurate modeling of these phenomena reqnires taking into account turbulence, heat transfer, fuel spray evaporation and detailed chemistry effects. In this contribution, numerical modeling based on Large Eddy Simulation (LES) is applied to investigate combustion processes of single component liquid spray jets.

Comprehensive reviews of LES combustion models in reacting single phase flows are provided in [18, 36, 37]. Extensive fundamental and applied researches were especially dedicated to address questions that govern the interacting phenomena in reactive multiphase flows. A recent review is provided by Sadiki et al. [40]. Pera et al. [32] and Zoby et al. [52] among others have proved a strong interdependence between combustion and disperse phase properties and highlighted the difficulty of isolating physical effects. Beside these studies, outstanding studies were also carried out in the modeling of spray ignition [45], [28], [49], reacting DNS [31, 45], [52], Conditional Moment Closure [24], [27], [26] DNS/LES coupling and transported filtered density function [14, 17, 19] of turbulent sprays. With respect to chemistry it appears that it is not realistic for engineering applications to solve transport eqnations for all species occurring in the chemical reaction process. Reduction techniques are often favored. Thereby one group is formed by the flamelet based tabulated chemistry along with the Flamelet Generated Manifold (FGM) (see e.g. [11, 47, 48, 50]) or the Flamelet Prolongated ILDM [13]. Though considerable efforts have been accomplished, applications of FGM based combustion modelling have not yet been done, to our knowledge, for spray combustion coupled to LES. Only recently Chrigui et al. [7] and Chrigui et al. [9] published their first achievements using LES to investigate spray jet flames of acetone and ethanol fuels. They focused on demonstrating the feasibility of classical LES coupled to an Eulerian-Lagrangian spray module to capture flow and combustion properties.

The present work aims at using this LES-based Eulerian-Lagrangian methodology to assess the impact of fuel and mass loading on the combustion properties of turbulent spray jets. Two different fuels, acetone and ethanol, are used to produce the spray jets through a pilot flame and a co-flowing atmospheric air. A spray pre-evaporation zone enables the combustion regime to turn from diffusion to partially premixed spray combustion. The methodology includes a full two-way coupling of the interacting two phases in presence, while the carrier phase turbulence is captured by the LES and the combustion by the FGM approach. The droplet evaporation is described by a non-eqnilibrium vaporization model along with a droplet Lagrangian tracking.

The paper is structured as follows. First the droplet Lagrangian tracking is introduced, followed by an outline of the non-eqnilibrium evaporation model. Then the modeling approach of LES completed by the FGM generation is highlighted. In section 3, the experimental configuration and the computational set up including the boundary conditions for both the carrier and the disperse phases are presented. Analysis, discussion and comparisons of the numerical results with the experimental data are provided in the subseqnent section while conclusions are summarized in the final section.

2 Modeling Approach

2.1 Disperse Phase Lagrangian Description

According to the Lagrangian approach, the eqnations of the droplet position x_{pi}, velocity u_{pi} and temperature T_p along the trajectory of each computational droplet in the carrier flow field have to be solved. Assuming a spherical, single parcel these eqnations are:

$$\frac{\mathrm{dx_{pi}}}{\mathrm{dt}} = \mathrm{u_{pi}}\,, \quad \mathrm{m_p}\frac{\mathrm{du_{pi}}}{\mathrm{dt}} = \mathrm{F}_i \tag{1}$$

$$\mathrm{m_p C_p}\frac{\mathrm{dT_p}}{\mathrm{dt}} = \mathrm{Q} \quad \text{and} \quad \frac{\mathrm{dm_p}}{\mathrm{dt}} = -\dot{\mathrm{m}}_{\mathrm{vap}} \tag{2}$$

where $m_p$is the mass of computational droplet or parcel, C_p the heat coefficient, F_i denotes the summation of all the forces acting on the parcel and Q the net rate heat transfer to the parcel while $\dot{m}_{vap}$ expresses the droplet vaporization rate, respectively. Since the ratio between the specific mass of liquid fuel and that of the gas phase mixture has a value around 10^3, we follow Chrigui et al. [7] and consider only the drag, gravitation and buoyancy forces to act on the droplet. Eq. (1) that describes the particle dynamics according to the Basset-Boussinesq-Oseen eqnation (BBO-eqnation) then reduces to:

$$\frac{du_{p,i}}{dt} = \frac{3}{4}\frac{C_W}{D_p}\frac{\rho}{\rho_p}\left|\vec{u} - \vec{u_p}\right|(u_i - u_{p,i}) + \frac{(\rho_p - \rho)}{\rho_p} g_i \tag{3}$$

The drag coefficient $C_W$is determined for a spherical, not deformable, droplet as proposed by Yuen and Chen (1976):

$$C_W = \frac{24}{\mathrm{Re}_p}\left(1 + \frac{1}{6}\mathrm{Re}_p^{2/3}\right) \qquad \mathrm{Re}_p \leq 1000 \tag{4}$$

$$C_W = 0.44\mathrm{Re}_p \geq 1000$$

where Re_p denotes the particle Reynolds number given by

$$\mathrm{Re}_p = \frac{D_p \left| u_i - u_{pi} \right|}{\nu}. \tag{5}$$

while the first term in (3) includes the particle-relaxation time, τ_d, expressed as

$$\tau_d = \frac{4 D_d \rho_d}{3 C_W \, \bar{\rho} \, |u_i - u_{pi}|} \tag{6}$$

Thereby D_d is the particle diameter, ρ_d the density of particle and ν the kinematic viscosity of the fluid.

It is worth mentioning that in Eq. (3) the flow velocity u_i appears in its instantaneous value.To quantify this instantaneous fluid velocity and its effect on the droplet distribution within the LES framework, the SGS values of the fluid parcel velocity at the droplet location should be modeled. As it is known from recent studies by Pozorski et al. [38] the impact of SGS dispersion can vary depending on the particle inertia parameter. In particular, for evaporating spray flow, the droplets become smaller and their inertia parameter changes, hence sooner or later the droplets unavoidably enter the size range where there is an impact from the flow SGS. In LES, reports from the literature highlighted the importance of the SGS in the prediction of the disperse phase properties (see in [7, 9, 19]). Though the SGS dispersion appears to be so relevant for the prediction of the reacting two phase flow, it is common practice in the Eulerian-Lagrangian LES studies of dispersed flows to neglect the SGS flow scales [2, 3, 19, 30, 43]. It is generally argued that the long-time particle dispersion is governed by the resolved, larger-scale fluid eddies. In this contribution the dispersion of droplet is not accounted for. We simply rely on the fact that at least 80 % of the instantaneous carrier phase turbulence level is captured by the resolved scales.

While writing Eq. (2) temperature variation inside the droplet is neglected and thus droplet temperature is considered uniform. This assumption is reasonable since dragged droplets have diameters in the range of 30 μm. Accordingly the Uniform Temperature (UT) model by Abramzon et al. [1] and Sirignano [46] is applied to describe the droplet evaporation process. This model describes the evolution of droplet's temperature and diameter, i.e. evaporation rate and energy flux through the liquid/gas interface. The non-eqnilibrium extension of this model is applied (see [25, 41]). Note that all the assumptions of this model are valid in the investigated configurations. In particular, break-up and coalescence are neglected to ensure that the evolution of the droplet diameter is only due to the evaporation processes. The Weber number (We) near the nozzle (x/D = 0.3), which is used as an indicator for the break-up phenomenon, is less than 0.3 in the configurations under study. The critical value, however, is about 40 times larger, i.e. $\mathrm{We}_{\mathrm{cri}} = 12.07$. The Ohnesorge Number (Oh) is less than 0.006 for all cases at the exit plane, therefore no further drop deformation and break up are possible downstream of the exit plane and thus the changes in the droplet size are due to evaporation only. Reviews of the evaporation models can be found in [1], [46], [4], [44] and [25].

2.2 LES Description

In the line of the FGM approach, the filtered transport eqnations for control variables, namely the mixture fraction and one reaction progress variable (RPV), are solved together with the filtered transport eqnations for mass density and momentum of the Newtonian fluid under investigation in a variable-density Low Mach number formulation as:

$$\frac{\partial \bar{\rho}}{\partial t} + \frac{\partial \bar{\rho}\tilde{u}_i}{\partial x_i} = \bar{S}_{vapor} \tag{7}$$

$$\begin{aligned}\frac{\partial}{\partial t}(\bar{\rho}\tilde{u}_i) + \frac{\partial}{\partial x_j}(\bar{\rho}\tilde{u}_i\tilde{u}_j) = -\frac{\partial \bar{p}}{\partial x_i} + \bar{\rho} g_i + \\ \frac{\partial}{\partial x_j}\left[\bar{\rho}\tilde{\nu}\left(\frac{\partial \tilde{u}_i}{\partial x_j} + \frac{\partial \tilde{u}_j}{\partial x_i}\right) - \frac{2}{3}\bar{\rho}\tilde{\nu}\frac{\partial \tilde{u}_k}{\partial x_k}\delta_{ij} - \bar{\rho}\tau_{ij}^{sgs}\right] + \bar{S}_{u,i}\end{aligned} \tag{8}$$

$$\frac{\partial}{\partial t}\bar{\rho}\tilde{z} + \frac{\partial}{\partial x_i}(\bar{\rho}\tilde{u}_i\tilde{z}) = \frac{\partial}{\partial x_i}\left(\bar{\rho}\tilde{D}_f\frac{\partial \tilde{z}}{\partial x_i}\right) - \frac{\partial}{\partial x_i}(\bar{\rho}J_i^{sgs}) + \bar{S}_{vapor} \tag{9}$$

$$\begin{aligned}\frac{\partial}{\partial t}\bar{\rho}\tilde{Y}_{RPV} + \frac{\partial}{\partial x_i}\left(\bar{\rho}\tilde{u}_i\tilde{Y}_{RPV}\right) = \frac{\partial}{\partial x_i}\left(\bar{\rho}\tilde{D}\frac{\partial \tilde{Y}_{RPV}}{\partial x_i}\right) \\ -\frac{\partial}{\partial x_i}\left(\bar{\rho}J_i^{sgs}\right) + \tilde{\dot{\omega}}_{RPV} + \tilde{\dot{\omega}}_{vapor}\end{aligned} \tag{10}$$

where the dependent filtered variables are obtained from spatial filtering, $\phi = \tilde{\phi} + \phi''$ with $\tilde{\phi} = \overline{\rho\phi}/\bar{\rho}$ and ϕ'' the subgrid scale (SGS) fluctuations. Thereby bars and tildes express mean and filtered quantities. In Eqs. (7)-(10) the variables u_i (i = 1, 2, 3) denote the velocity components at x_i direction, ρ the density, p the hydrostatic pressure and δ_{ij} the Kronecker delta. The quantity ν is the kinematic molecular viscosity and D_f the molecular diffusivity coefficient. Following Chrigui et al. [7] the mixture fraction, z, is defined according to Bilger et al. [5] as:

$$z = \frac{a - a_{Oxidizer}}{a_{Fuel} - a_{Oxidizer}} \tag{11}$$

where

$$a = 2\frac{Y_C}{M_{w,C}} + 0.5\frac{Y_H}{M_{w,H}} - \frac{Y_O}{M_{w,O}}. \tag{12}$$

The parameters Y_C, Y_H and Y_O denote the element mass fractions of carbon, hydrogen and oxygen atoms. $M_{w,C}$, $M_{w,H}$ and $M_{w,O}$ are the molecular weights. At the inlet the values of the oxidizer and fuel are given by $a_{Oxidizer}$ and a_{Fuel}, respectively.

Table 1 Source terms due to the presence of droplets

ψ	$\tilde{S}_{\psi,i}$
1	$\sum_p \frac{\dot{m}_{p,v} N_p}{V_{i,j,k}}$
$\tilde{u}$	$-\sum_p \frac{\dot{m}_p N_p}{V_{i,j,k}} [(u_p^{t_n+\Delta t}) - u_p^{t_n}) - g_x \Delta t] + \sum_p \frac{\dot{m}_{p,v} N_p}{V_{i,j,k}} u_p^{t_n}$
$\tilde{v}$	$-\sum_p \frac{\dot{m}_p N_p}{V_{i,j,k}} [(v_p^{t_n+\Delta t}) - v_p^{t_n}) - g_y \Delta t] + \sum_p \frac{\dot{m}_{p,v} N_p}{V_{i,j,k}} v_p^{t_n}$
$\tilde{w}$	$-\sum_p \frac{\dot{m}_p N_p}{V_{i,j,k}} [(w_p^{t_n+\Delta t}) - w_p^{t_n}) - g_z \Delta t] + \sum_p \frac{\dot{m}_{p,v} N_p}{V_{i,j,k}} w_p^{t_n}$
$\tilde{z}$	$\sum_p \frac{\dot{m}_{p,v} N_p}{V_{i,j,k}}$

The quantity $\tilde{Y}_{RPV}$ is the filtered concentration of the reaction progress variable and D denotes the molecular diffusivity coefficient. The Eqs. (7)-(10) govern the evolution of the large, energy-carrying, scales of flow and scalar field. The effect of the small scales in the flow and scalar field appears through the SGS stress tensor and the SGS scalar flux vector

$$\tau_{ij}^{SGS} = u_i u_j - u_i u_j \tag{13}$$

$$J_i^{sgs} = u_i \xi - u_i \xi, \ \xi \equiv (z, Y_{RPV}) \tag{14}$$

respectively. The SGS stress tensor is postulated by a Smagorinsky-model with dynamic procedure according to Germano et al. [15]. In order to stabilize the model, the modification proposed by Sagaut [42] is applied. In addition a clipping approach will reset negative Germano coefficient C_s to zero to avoid destabilizing values of the model coefficient. It is known that wall-adaptive SGS models have been proposed recently, like wall adapting laminar eddy (WALE) model or the Vreman model with and without dynamic procedure, see [42]. Nevertheless, no special wall-treatment is included in the SGS model. We rather rely on the ability of the dynamic procedure to capture the correct asymptotic behavior of the turbulent flow when approaching the wall [51]. To represent the SGS scalar flux in the mixture fraction and in the RPV eqnations a gradient ansatz (15) is used with a constant turbulent Schmidt number of 0.7.

$$J_i^{sgs} = -\frac{\upsilon_t}{\sigma_t} \frac{\partial \tilde{\xi}}{\partial x_i}; \upsilon_t = C_s \Delta^2 \left| \overline{S} \right| \tag{15}$$

υ_t is SGS viscosity, σ_tSGS Schmidt number and $\left|\overline{S}\right|$the absolute values of strain rate.

The source terms $\overline{S}_{u,i}$ and $\overline{S}_{vapor}$ that characterize the direct interaction of mass, momentum, and mixture fraction between the droplets and the carrier gas are summarized in Table 1. The variable ψ represents the mass density, velocity components (u, v, w) and the mixture fraction, respectively. The quantities, $\tilde{u}$, $\tilde{v}$ and $\tilde{w}$, are the filtered gas phase (axial, tangential and transversal) velocity components while u_p, v_p and w_p represent the corresponding velocity components of the parcel. N_p is the number of real droplets represented by one numerical droplet and V_{ijk} the cell volume. The quantity g represents the gravitation, $\dot{m}_p$ the amount of mass released by a parcel when it crosses a control volume (CV) per second while Δt stands for the Lagrangian integration time step.

Concentrating on Eq. (10), notice that the term, $\tilde{\dot{\omega}}_{RPV}$, represents the classical filtered chemical reaction rate [13, 18, 35, 36, 37, 33, 48] and the last term accounts for the evaporation contributions into the RPV eqnation [32]. Some other additional terms may emerge in a general expression of the RPV eqnation that explicitly includes the effect of evaporation on combustion [32]. Assuming that all droplets have evaporated before combustion, only the quantity $\tilde{\dot{\omega}}_{RPV}$ needs further modeling within the FGM approach. In the present work the RPV is defined as

$$Y_{RPV} = \frac{Y_{CO_2}}{M_{CO_2}} + \frac{Y_{H_2O}}{M_{H_2O}} + \frac{Y_{H_2}}{M_{H_2}}, \tag{16}$$

where Y_{CO_2}, Y_{H_2O} and Y_{H_2} are the mass fraction of CO_2, H_2O and H_2, respectively. The quantity M represents the molar mass, so its reciprocal is used as a weighting factor for each species. As pointed out elsewhere, the choice of the species defining the RPV depends on the problem being solved. In this specific case, the three major species retained are considered to properly capture the reaction zone. Using the two parameters (z, Y_{RPV}) a two-dimensional manifold is then generated by means of the CHEM1D code [6] by simply simulating a set of 1D diffusion flamelets [6, 13, 20, 33, 35, 48, 50] with increasing scalar dissipation rate, and thereafter switching to unsteady flamelets when reaching the critical scalar dissipation rate.

The filtered combustion variables required in the LES are then retrieved by integrating over the joint PDF of the mixture fraction and the defined RPV:

$$\tilde{\phi} = \int_0^{\infty} \int_0^{\infty} \phi(z, y) P(z, y) dz\, dy; \quad \mathrm{y} \equiv \mathrm{Y}_{RPV}. \tag{17}$$

where, $P(z, y)$ is the joint PDF. It is practical to carry out the pre-integration upon the normalized values, where the RPV is normalized by its maximum value restricting the integration domain to lie in [0, 25]. Assuming a statistical independence between the RPV and the mixture fraction yields:

$$\tilde{\phi} = \int_0^{1} \int_0^{1} \phi(z, y) \tilde{P}(z) \tilde{P}(y^*) dz\, dy^* \tag{18}$$

where the Favre-weighted PDF is derived from a standard one, $\widetilde{P}\,(y^*) = \rho\,(y^*)\,P\,(y^*) \big/ \bar{\rho}$, and y^* the normalized RPV. Since the mixture fraction is no more a conservative quantity, it may influence the PDF distributions. Gutheil et al. [14, 17] showed from a comparison of Monte-Carlo PDF with standard *beta*-PDF that a *beta*-function describes the actual shape of the PDF differently. Nevertheless a presumed *beta*-PDF distribution is chosen here as crude approximation. This implies the mixture fraction depends on its first and second moments. Effects of this assumption on predicted RANS results were reported in [20]. In LES, these have not been investigated in the literature yet.

In Eq. (18) the PDF of the RPV needs to be estimated. As a first-order approach, the δ-function is applied, allowing the combustion variables to be function of the

RPV mean values only. This assumption implies that the fluctuations of the RPV are sufficiently resolved or they could be omitted. This is realistic for spray flames under study, since they tend to exhibit diffusion flame behavior in which the RPV fluctuations are not large compared to premixed cases. The joint PDF in (2, 17) yields

$$P(z, y^*) = P(z) \cdot P(y^*) = \beta(z; \tilde{z}, z''^2) \cdot \delta(y^*) = P(z; \tilde{z}, z'',^2 y^*). \tag{19}$$

Thermo-chemical quantities in Eq. (2, 17) can then be parameterized and tabulated in so-called pre-integrated tables (tabulated SGS chemical parameters) as function of filtered mixture fraction, its variance and normalized filtered RPV as:

$$\tilde{\phi} = f(\tilde{z}, z'',^2 y^*). \tag{20}$$

While generating the FGM table, the effect of droplet evaporation along with the interaction between evaporating droplets and combustion is not directly accounted for. To do this, at least the vaporized mass quantity has to be included as parameter in Eq. (20). This work is still in progress.

The reliability of spray combustion models depends primarily on how the fuel-air mixture preparation is accurately described. It is then of interest to better understand the behavior of vapor concentration along with the mixture fraction variance when spray evaporation is present. To this purpose, a transport eqnation of mixture fraction variance is suitable [31, 32, 39]. Sadiki et al. [39] investigated the impact of the modeling of the evaporation source term in the transport eqnation of mixture fraction variance on the prediction of combustion properties in RANS context. Thereby the outcomes of two models have been compared. It turned out that the model by Reveillon et al. [31] could deliver more accurate prediction in comparison to the formulation by Hollman et al. (see in [39]). Because the proper contribution of the evaporation source term in the eqnation of the RPV (see Eq. (10)) has been neglected as complete evaporation has been assumed before combustion, the mixture fraction variance is obtained simply following the algebraic gradient formulation [7, 35]

$$z''^2 \approx {C_{eq}}^2 \left(\frac{\partial \tilde{z}}{\partial x_i} \right)^2 \tag{21}$$

The model coefficient, C_{eq}, lies into [0.1, 0.2] and is set to 0.15 in the present work.

The combustion of acetone is modeled by a detailed reaction mechanism including 84 species and 409 elementary reactions as developed in [34]. Ethanol is modeled by means of a detailed chemical reaction mechanism as developed and validated by Marinov [22]. It consists of 56 species and 351 reactions.

3 Investigated Configurations and Numerical Set Up

3.1 Experimental Configuration and Computational Sset Up

The configuration used to study the ethanol and acetone spray combustion is shown in Fig. 1. It represents the setup experimentally investigated by Masri and Gounder

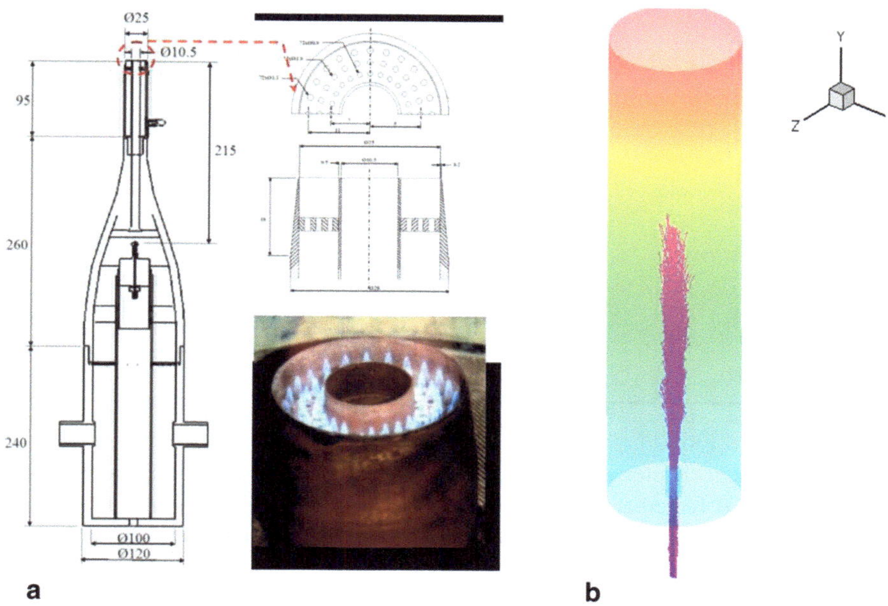

Fig. 1 **a** Schematic of the spray [40] burner set up **b** Computational domain

[16, 23]. The burner is mounted vertically in a wind tunnel that supplies a co-flowing air stream of 4.5 m/s. The co-flow of diameter 104 mm surrounds the burner. The contraction of the carrier phase topology has a ratio of 10:1. The outer diameter of the annulus is 25.0 mm whereas the lip thickness is 0.2 mm. The pilot flame that is set to a stoichiometric mixture of hydrogen, acetylene and air has an un-burnt bulk velocity of 1.5 m/s. Its border is mounted 7 mm upstream of the nozzle exit and contains 72 holes. The co-flow and nozzle exit plane is 59.0 mm downstream of the tunnel exit plane. This provides an unconfined working section.

The spray is initialized 215 mm upstream of the nozzle exit plane and exhibits a poly-disperse behavior after traveling a pre-vaporization zone in which small classes evaporate before reaching the exit of the nozzle. The resulting ethanol and acetone flames feature a partially premixed character. A detailed description of the experimental setup and apparatus used for the generation of the experimental data is provided by Masri and Gounder [23] (see also [16, 47]).

In [26] the non-reacting and reacting cases SP1, SP2 and EtF1, EtF4 and EtF7 have been investigated using RANS and CMC. Applying RANS and FGM, Chrigui et al. [8] studied the configurations SP1/AcF1, SP2/AcF2 and SP5/AcF5. Using LES, Chrigui et al. analyzed recently the evaporation process of the configurations EthF1, EthF3, EthF6 and EthF8 (see in this book) and the spray combustion of AcF3, AcF6 and AcF8 [7]. They kept the same liquid fuel injection rate while varying the jet Reynolds number as well as the carrier mass flow rate along with the spray jet density. They also reported LES results of the configurations EtF3 and EtF8 in [9, 40].

Table 2 Different test cases with the flow boundary conditions using both fuels, acetone and ethanol, respectively

Acetone Combustion	*Flame test case AcF3*	*Flame test case AcF6*
$\dot{m}_l[g/\min]$	45	45
$U_{jet}[m/s]$	24	36
$\dot{m}_{air}[g/\min]$	150	225
$\mathrm{Re}_{jet}[-]$	20730	28076
$\phi_{golbal}[-]$	2.9	1.9
Ethanol Combustion	*Flame test case EtF3*	*Flame test case EtF6*
$\dot{m}_l[g/\min]$	45	45
$U_{jet}[m/s]$	24	36
$\dot{m}_{air}[g/\min]$	150	225
$\mathrm{Re}_{jet}[-]$	19678	28076
$\phi_{golbal}[-]$	2.9	1.9

In the present chapter, a LES-based study is carried out in which the configurations EtF3 and EtF6 fueled with ethanol as well as AcF3 and AcF6 using acetone are compared in terms of fuel influence and mass loading impact on the combustion properties.

3.2 Boundary Conditions

All the boundary conditions for the carrier phase are provided in Table 2. A decreasing mass loading in the inner jet from 30 to 15 % could be calculated. The velocity components of the carrier phase are given as block profile at the inlets and the Reynolds numbers from Table 2 attest a highly turbulent two phase flow. As the carrier phase travels a distance 20D to reach the nozzle exit plane, the flow develops turbulent structures, even with block velocity profiles.

Following Chrigui et al. [7, 9, 40] the configuration under study was numerically represented by a computational domain consisting of 17 blocks that count 1.1×10^6 control volumes (cv), Fig. 1b. Within one coupling time step the number of parcels injected is 2500 while the number of time steps achieved between both phases, that represent the fluid data and/or source term transfer, exceeds 320,000 couplings. The averaging of the spray properties is thus performed over more than 750×10^6 parcels. The disperse phase properties are statistically independent and not conditioned on the number of parcels tracked or coupling time steps. The TVD scheme is applied for the velocity exit boundary with a condition with 6 m/s. For the RPV the boundary condition is set to zero in the entire domain except at the pilot flame inlet, where it is set to the maximum absolute value that eqnals 0.0101. Note that the total number of the tracked parcels exceeded 1 million within one coupling-iteration.

Since the spray is injected by an ultrasonic nebulizer, the droplet size distribution produced is known to be approximately lognormal. Close to the jet wall the measured droplet size PDF shows a bias towards small droplets. According to Chrigui et al.

[7, 9, 40] the drop deformation regime map as a function of We and *Oh*, provided by Faeth [12], was used for determining if further droplet deformation or break up occurred in the spray jets. Here the Weber number was found to be less than 0.3 and Oh was less than 0.006 for droplets in the sizes range $40 < D_p < 50\ \mu$. Thus the droplets are assumed spherical and will not undergo any further deformation due to droplet break up, downstream of the nozzle exit plane. The simulations are performed using 12 different classes of droplets. It is remarkable that almost all classes possess the same injection axial velocity that eqnals 42 m/s, whereas the standard deviation corresponds to ca. 3 m/s yielding an axial turbulence intensity of 7.5 %.

3.3 Numerical Implementation

According to Chrigui et al. [7, 9, 40] the governing eqnations of the carrier gas phase are discretized in the 3D low-Mach number LES code FASTEST. The code is able to compute 3D-complex geometries by using flexible, block structured and boundary fitted grids [21, 29]. The code uses the finite volume method in which a co-located grid is applied. For spatial discretization, specialized central differencing schemes that hold the second order for arbitrary grid cells are used [21]. The convective term in the scalar transport is discretized using non-oscillatory, bounded TVD schemes. For the time stepping, multiple stage Runge-Kutta schemes with second order accuracy are used. A fractional step formulation is applied and at each stage a momentum correction is carried out in order to satisfy continuity. The parcels are tracked using LAG3D code in which the eqnation of motion, the temperature evolution and the evaporation rate are discretized using Euler first/second order schemes and solved explicitly [7–10, 39, 40].

4 Results and Discussions

Let us first compare the reference flames EtF3 and AcF3 using both the ethanol and the acetone fuel, in Fig. 2. Here the instantaneous contours of the OH mass fraction are plotted in the cross section along the center line for EtF3 and AcF3. The maximum value is obviously registered at the reaction zone where mixture fraction is close to stoichiometry. Downstream of the nozzle exit and at the centerline of the configuration, no flame is observed, probably caused by the high strain rate at the nozzle exit plane or by a lack of sufficient heat to maintain the combustion at the centerline. Further downstream, at $y = 0.3D$, due to momentum transfer from axial to radial direction, the OH reaches the maximum value. At the centerline and downstream the nozzle the temperature should reduce its value below the inlet boundary condition because of spray evaporation. This heat loss effect is not considered in the modeling yet. Depending on the evaporation rate the temperature plot is changing. If droplets evaporate within the pre-vaporization zone completely, the combustion regime will be likely to demonstrate a premixed nature. Note that the flammability

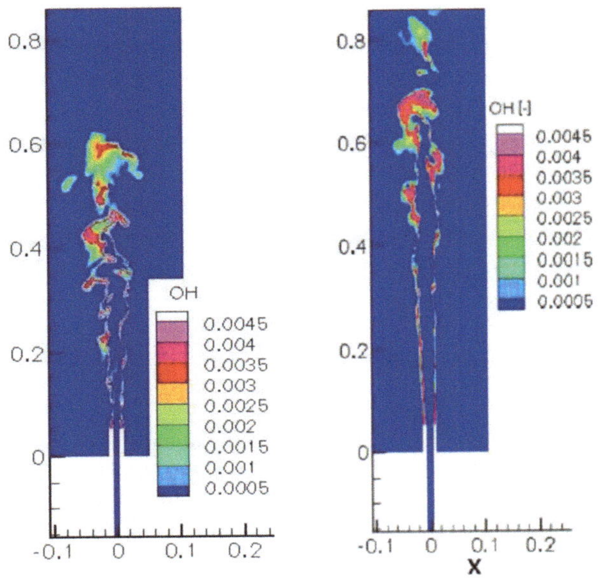

Fig. 2 Instantaneous plots of the OH mass fraction of the ethanol spray flame (*left*) and acetone flame (*right*) along the axial direction

limit of the ethanol lies between 0.44 and 3.34 while it is between 0.64 and 3.59 for the acetone. For very slow vaporization, i.e. most droplets exit the nozzle without evaporation; the spray flame is likely to have a diffusion flame behavior. Worth mentioning is that, numerically, the combustion cannot be sustained without setting the RPV boundary condition at the pilot flame location to its maximum value. An initialization of the RPV at the nozzle exit plane is not enough to make the mixture burn, because the RPV is transported downstream and the mixture blows out. In order to stabilize the flame, an initialization reqnires a recirculation zone which is not present in this configuration. The OH mass fraction, which is an indicator of the position of the flame, varies between "0" and 4.5×10^{-3} in both flames. However, the acetone flame appears higher than the ethanol one (see also Figs. 7 and 8).

Figures 3, 4, 5 and 6 show the axial droplet velocities and corresponding fluctuations of all the cases under investigation. Reasonable agreement for the mean droplet velocities is observed in the first cross-section. At x/D = 20 and x/D = 30, small discrepancies are observed in the averaged droplet velocity in the ethanol case. Unfortunately, a comparison between simulated gas phase velocity (that may help to clarify these discrepancies) and experimental data of the carrier phase (that are not available) is not possible. Disagreement at the last cross-sections may originate from the presence of remaining big droplets which are not following the carrier phase.

As the gas phase is captured using LES, mean velocities of droplets are well predicted. The axial velocity of the acetone test cases shows, at x/D = 30, an increase of its value compared to ethanol. Acetone is more volatile and has a lower boiling point than ethanol, thus it evaporates faster. The added vapor mass within a control volume accelerates the carrier phase, which drags the disperse phase faster, causing thus the

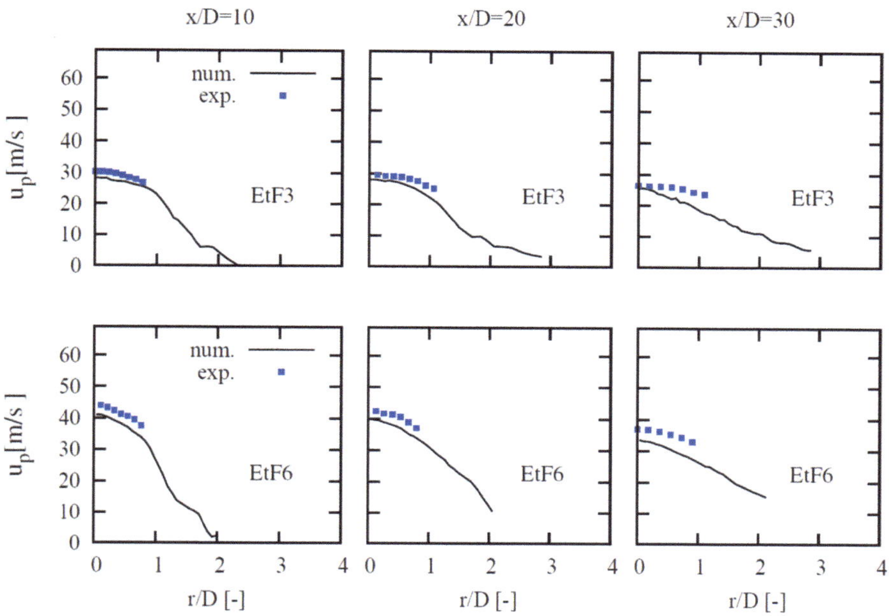

Fig. 3 Radial profiles of the droplet mean axial velocity at different distances from the nozzle exit plane for ethanol case. The x-axis represents r/D

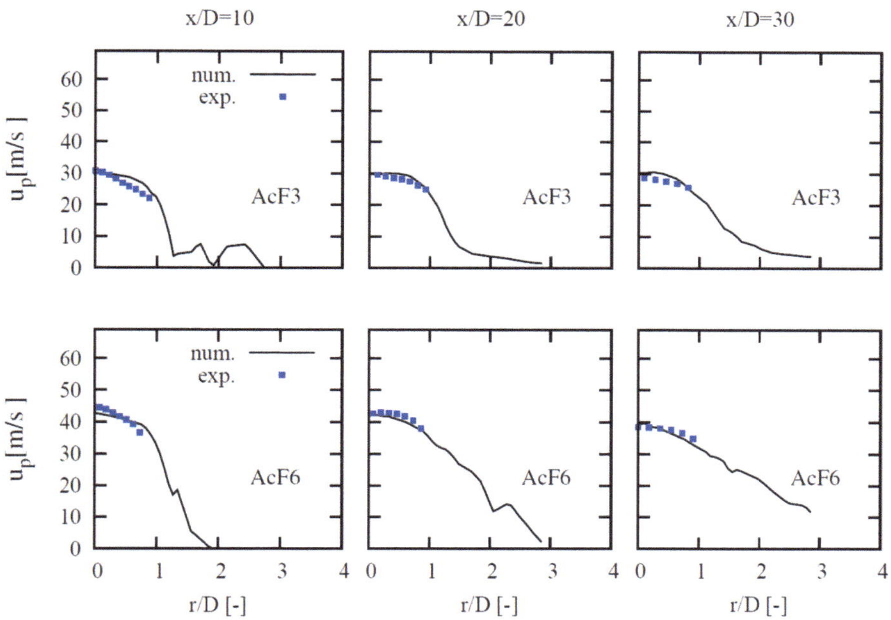

Fig. 4 Radial profiles of the droplet mean axial velocity at different distances from the nozzle exit plane for acetone case. The x-axis represents r/D

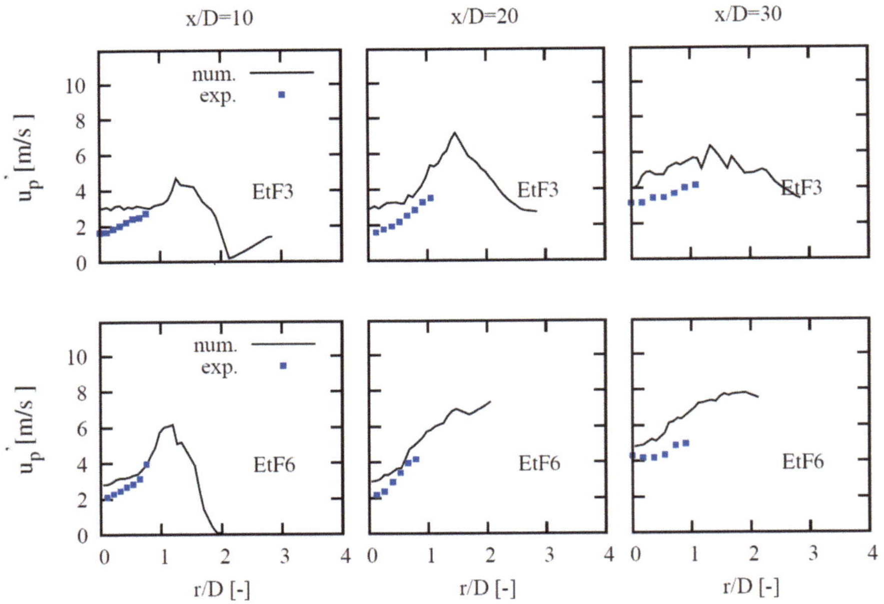

Fig. 5 Radial profiles of the droplet axial velocity fluctuation at different distances from the nozzle exit plane for ethanol. The x-axis represents r/D

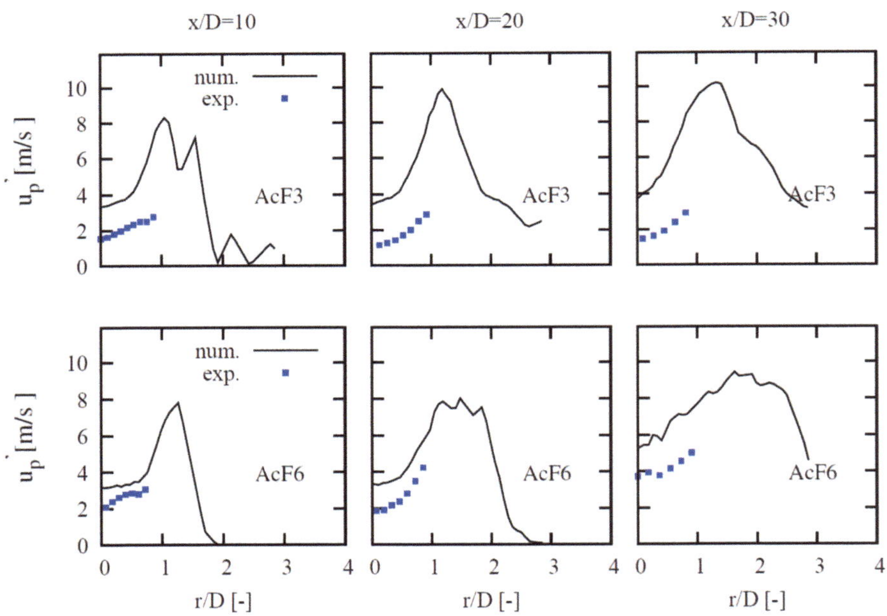

Fig. 6 Radial profiles of the droplet axial velocity fluctuation at different distances from the nozzle exit plane for acetone. The x-axis represents r/D

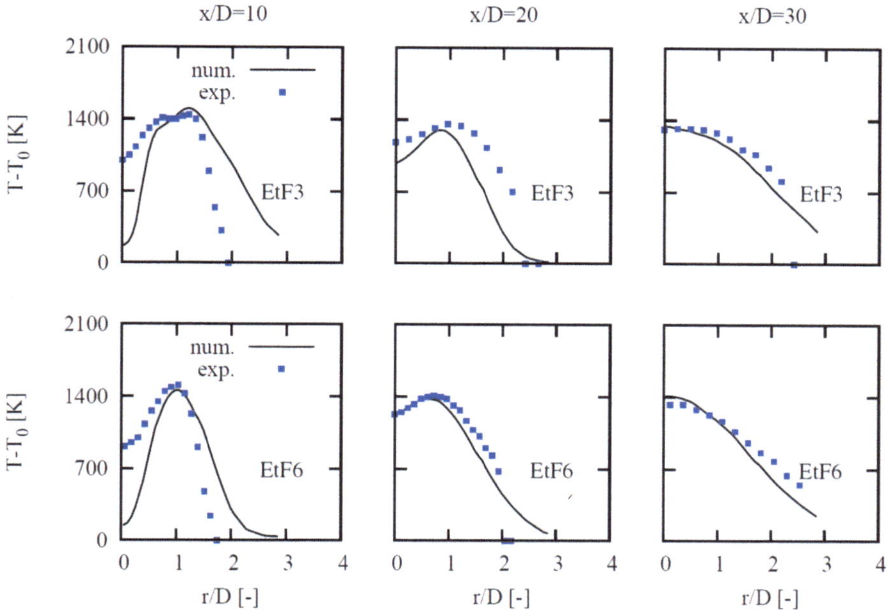

Fig. 7 Radial profiles of the excess temperature at different distances from the nozzle exit plane for ethanol. The x-axis represents r/D

increase of the droplet mean velocities. Worth noting is that the measurements are restricted to r/D = 1 since the number of droplets outside this region is not important and delivers no accurate statistics. The fluctuations of the droplet velocities of both fuels are comparable. The effect of faster evaporation is not remarkable. Numerically, acetone droplets show a higher RMS, particularly at the shear flow region, i.e. between the central jet and the co-flow. The difference between both fuels in the droplet velocity fluctuation is most likely caused by the difference in the droplet diameters. As acetone evaporates faster, corresponding droplets are more sensitive to the turbulence. Acetone droplets rather tend to follow the carrier phase dynamics than ethanol droplets, thus their RMS gets increased because of the important velocity gradient between r/D = 1 and r/D = 2.

The discrepancies between the experimental and numerical results in the droplet velocity fluctuations, observed in Figs. 5 and 6, could be caused by the spray-wall interaction, which is not modeled. At the nozzle exit, droplets are observed accumulating in the region close to the nozzle edge. A dense region is formed, thus collision and/or coalescence as well as breakup phenomena are likely to take place. As result, droplets are highly disturbed; they change their velocity and direction. Hence the turbulence level of the dispersed phase increases. The neglect of the effect of SGS on the dispersion of the spray may be another source of the discrepancies in the droplet fluctuations. Indeed small particles, as it is the general case in evaporating droplets, tend to follow the carrier phase dynamics, which is captured by the resolved part and the SGS contribution. This SGS dispersion is unfortunately not included yet.

The temperature plot in Figs. 7 and 8 is a function of the RPV, mixture fraction and corresponding variances. Using these variables as input parameters to the FGM

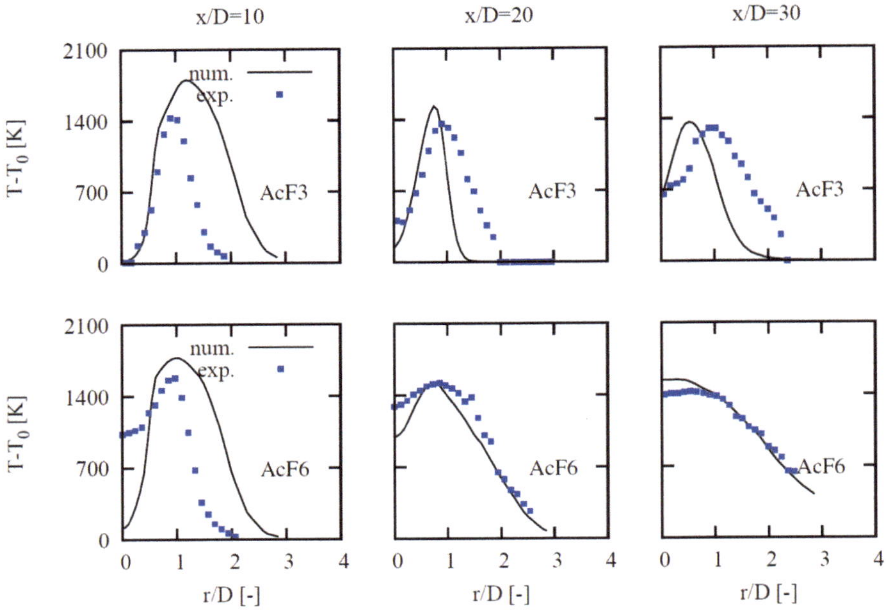

Fig. 8 Radial profiles of the excess temperature at different distances from the nozzle exit plane for ethanol. The x-axis represents r/D

table, the temperature is determined at each cell of the computational domain. It varies between the temperature of pure mixing and the maximum in case of eqnilibrium combustion. It is possible for the highest value of the RPV that different temperatures could be observed, depending on the mixture fraction and its variance. The center line of the configuration shows a cold flow despite the fact that the mixture fraction is close to stoichiometry.

Ethanol flames are broader and demonstrate an important value at the center line, i.e. $r/D = 0$. The mixture fraction distribution is altered by the rate of vaporization. The difference on the temperature profiles may also originate from the reaction mechanism, which involves different species and reactions number as well as laminar burning velocities. The numerical simulations show that the prediction of the ethanol flame agrees most favorably with the experimental data.

Figures 9 and 10 show the droplet volume flux at different axial cross- sections for ethanol and acetone, respectively. The numerical simulations predict the volume flux well and show reasonable agreements with the measurements. The slight over-prediction is expected, since not all droplets that cross a section could be experimentally captured, validated and registered. The real and correct profile of the droplet volume flux should be higher than the experimental curves provide. A comparison between the plots in Figs. 9 and 10 shows that the volume fluxes of acetone is smaller than that of ethanol. This indicates that the evaporation of the ethanol is slower than that of acetone.

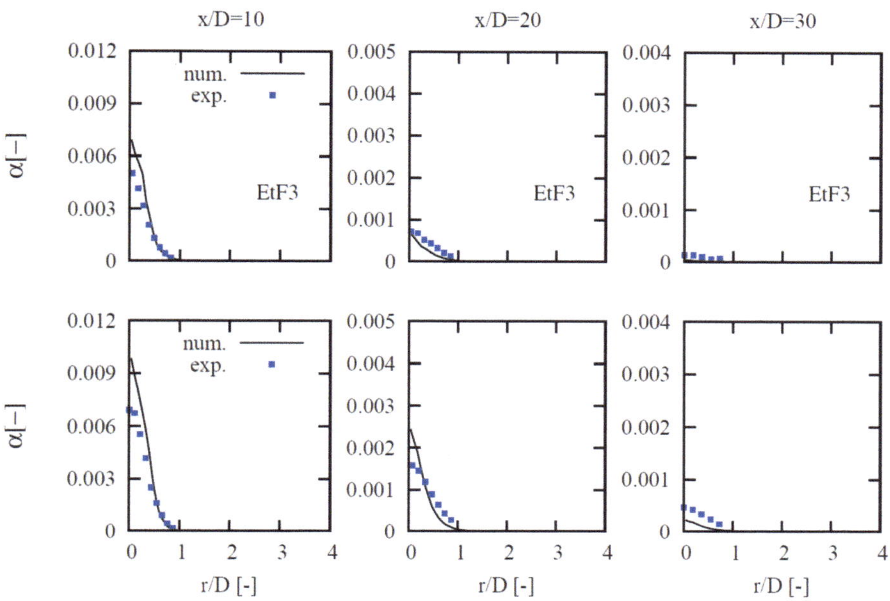

Fig. 9 Radial profiles of the droplet volume flux at different distances from the nozzle exit plane for ethanol. The x-axis represents r/D

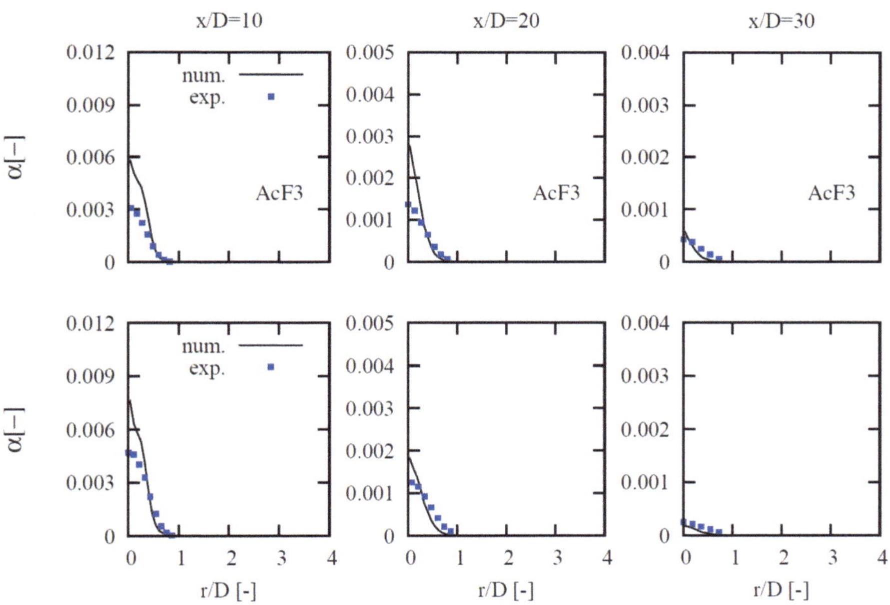

Fig. 10 Radial profiles of the droplet volume flux at different distances from the nozzle exit plane for acetone. The x-axis represents r/D

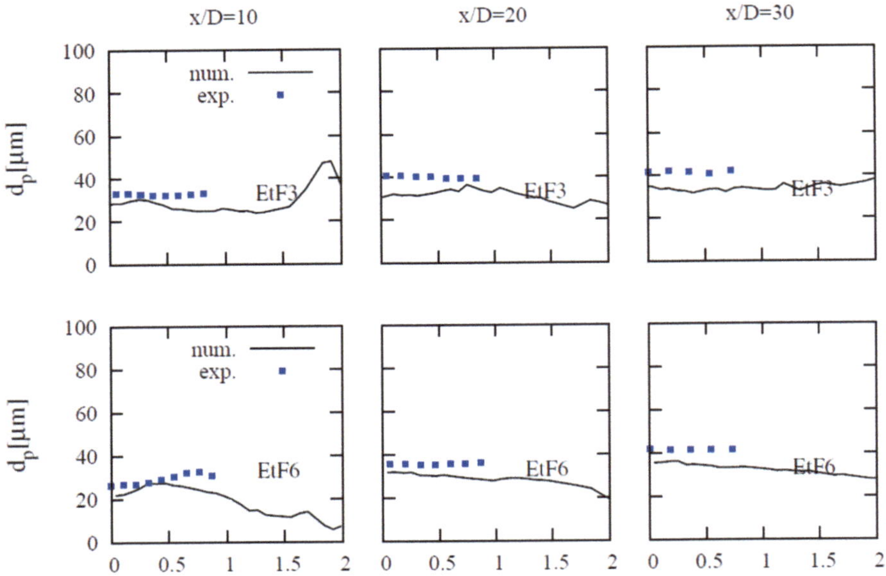

Fig. 11 Radial profiles of the droplet mean diameter at different distances from the nozzle exit plane for ethanol. The x-axis represents r/D

Figures 11 and 12 display the droplet mean-diameters for the ethanol and acetone spray at different cross-sections, respectively. A good agreement between the experimental data and the numerical simulation for all test cases is observed. Both sprays show similar behavior and evaporate in a comparable way. The acetone flame shows at $x/D = 20$ a lower diameter profile compared to ethanol which confirm that the acetone evaporates slightly faster.

Though the dispersed phase includes 12 different classes and demonstrates a polydisperse flow, the spray mean-diameters show nearly constant uniform profiles. This effect may be caused by the quick vaporization of the small classes that includes small droplet diameters, i.e. below 20 μm. Larger droplets, however, evaporate slower, they are dragged from the injection location to the nozzle exit within a highly turbulent flow, which increases the homogeneity of the class distribution in the prevaporization zone. Thus the spray mean diameter shows an approximately uniform profile in the radial direction.

5 Conclusion

An LES based investigation was carried out to analyze and evaluate the impact of fuel and mass loading on turbulent spray combustion. To retrieve the flow, mixing and combustion properties, an Eulerian-Lagrangian approach was adopted. The method included a full two-way coupling between the interacting two phases in presence, while the evaporation process was described by a non-eqnilibrium vaporization model. The carrier phase turbulence was captured by a combustion LES technique.

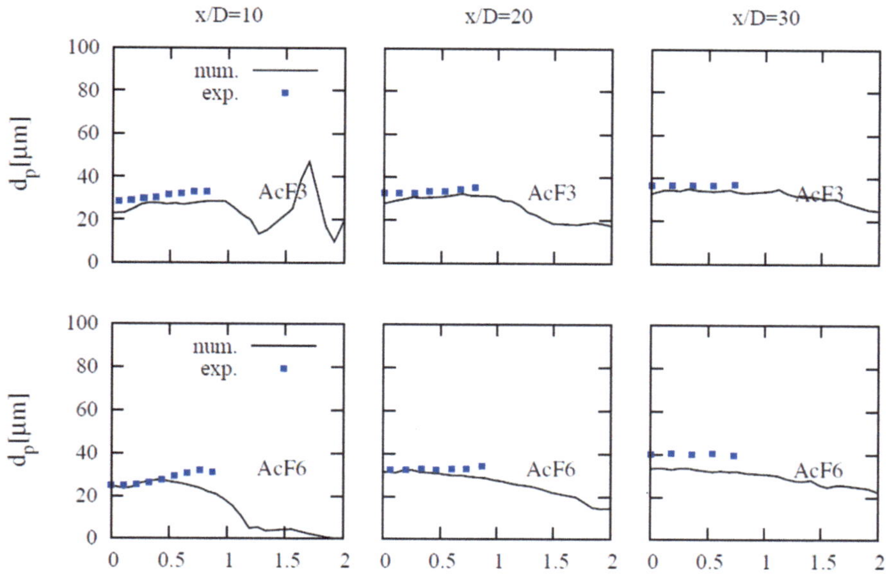

Fig. 12 Radial profiles of the droplet mean diameter at different distances from the nozzle exit plane for acetone. The x-axis represents r/D

In both cases, the ethanol and acetone fueled configurations, the occurring flow and combustion properties were numerically analyzed and compared with experimental data under different mass loading conditions. The impact of fuel and mass loading on turbulent spray combustion was evaluated in terms of flame structure, exhaust gas temperature, droplet diameters and velocities, droplet velocity fluctuations, and spray volume flux at different distances from the exit planes. The following conclusions can be drawn:

1. The axial droplet velocities predicted by simulations agree well with the experimental data. The discrepancies of the corresponding fluctuations can be attributed to the spray-wall interaction at the nozzle exit or/and to the deficiency of capturing the resolved instantaneous fluid velocity. This shows the necessity of a SGS dispersion model for evaporating droplets.
2. Acetone, which has a lower boiling point than ethanol, evaporates faster. This effect increases the carrier phase velocity, which in turn increases the drag of the spray, making thus the droplets move faster. The rate of vaporization influences the formation of the fuel/air mixture. The resulting mixture fraction of acetone/air demonstrates a richer character at the pre-vaporization zone and tends to a premixed behavior rather than the ethanol test cases.
3. The spray volume flow rates are in acceptable agreement. Thereby the simulation reveals the importance of polydispersity and accounts better for the actual amount of droplets than the experimental measurements.
4. With respect to spray combustion properties, the FGM concept captured well the flame height and lift-off. Although the effect of droplets on the thermocouple is

not quantified, spray evaporation and/or single droplet combustion at the thermocouple surface can produce large errors (up to 20 %). The temperature profiles demonstrated a fair agreement with the measurements.

In order to improve the results of the temperature, heat losses due to radiation and heat sinks because of spray evaporation should be considered.

Further improvements are suitable. The group combustion, that influences the spray combustion, should not be disregarded in the modeling. Droplet-droplet interaction along with the spray wall interaction should be added to improve the prediction of droplet distribution at the exit plane. An accurate SGS dispersion model needs to be integrated. It is also of great interest to generate the FGM table under consideration of the droplet vaporization process.

Acknowledgements The financial support by the Deutsche Forschungsgemeinschaft (DFG) is highly recognized.

References

1. B. Abramzon and W. A. Sirignano, Droplet Vaporization Model for Spray Combustion Calculations, Int. J. Heat Mass Transfer, Vol. 32, 1989, pp. 1605–1618.
2. S. V. Apte, K.Mahesh, M. Gorokhovski, P. Moin, Stochastic modeling of atomizing spray in a complex swirl injector using large eddy simulation, Proceedings of the Combustion Institute, Volume 32, Issue 2, 2009, Pages 2257–2266.
3. J. Bellan, L.C. Selle, Large Eddy Simulation composition eqnations for single-phase and two-phase fully multicomponent flows Original, Proceedings of the Combustion Institute, Volume 32, Issue 2, 2009, Pages 2239–2246.
4. A. Berlemont M. S. Grancher and G. Gouesbet, Heat and mass transfer coupling between vaporizing droplets and turbulence using a Lagrangian approach, J. of. Heat and Mass Transfer Vol.38, 1995, 3023–3034.
5. R. W., Bilger, S. H., St°arner, and R. J., Kee, On reduced mechanisms for methane-air combustion in nonpremixed flames. Combustion and Flame, 80, 1990, pp 135–149.
6. CHEM1D. http://www.tue.nl/en/university/departments/mechanical-engineering/research/research-groups/combustion-technology/flamecodes/chem1d/.
7. M. Chrigui, J. Gounder, A. Sadiki, A. R. Masri, J. Janicka, Partially premixed reacting acetone spray using LES and FGM tabulated chemistry, Combustion and Flame, Volume 159, Issue 8, Pages 2718–2741, August 2012.
8. M. Chrigui, J. Gounder, A. Sadiki, A. R. Masri, J. Janicka, Acetone Droplet Behavior in Reacting and Non Reacting Turbulent Flow, Flow Turbulence and Combustion, Volume 90, Issue 2, pp 419–447, 2013.
9. M. Chrigui, A. R. Masri, Amsini Sadiki, Johannes Janicka, Large Eddy Simulation of a Polydisperse Ethanol Spray Flame, Flow Turbulence and Combustion, Volume 90, Issue 4, pp 813–832, June 2013.
10. M. Chrigui, F. Sacomano, A. Sadiki A. R. Masri, Evaporation Modeling for Polydisperse Spray in Turbulent Flow, (present book), TCS3-Book-chapter.
11. U. Eguz, L.M.T. Somers, L.P.H. de Goey, Modeling of PCCI Combustion with the FGM approach, 13th International Conference on Numerical Combustion April 27–29, 2011, Corfu, Greece.
12. G. M. Faeth, Spray combustion phenomena, Proc. Combust. Inst. 26,1996, pp 1593–1612.

13. B. Fiorina, O. Gicquel, L. Vervisch, S. carpentier, N. Darabiha: Premixed turbulent combustion modeling using tabulated detailed chemistry and PDF, Proc. of the combustion institute, Vol. 30, pp. 867–874 (2005).
14. H.W. Ge and E. Gutheil, Probability density function (pdf) simulation of turbulent spray flows, Atomiz. Sprays 16 (2006), pp. 531–542.
15. M. Germano, U. Piomelli, P. Moin, and W. H. Cabot, A dynamic sub-grid scale eddy viscosity model, 1991, Phys. Fluids A, 3: 1760–1765.
16. J. D. Gounder and A. R. Masri, Flow field and Mass Flux Measurements near the Exit Plane of Spray Jets, ICLASS 2009, 11th Triennial International Annual Conference on Liquid Atomization and Spray Systems, Vail, Colorado USA, July 2009.
17. E. Gutheil, Modeling and Simulation of Droplet and Spray Combustion, Handbook of combustion, Wiley-VCH Verlag GmbH & Co. KGaA, 2010.
18. J. Janicka, A. Sadiki: Large Eddy Simulation of turbulent combustion systems, Proc. Of the combustion institute, 30, pp. 537–547, 2005.
19. W.P. Jones, S. Lyra, A.J. Marquis, Large Eddy Simulation of evaporating kerosene and acetone sprays, International Journal of Heat and Mass Transfer, Vol. 53, Issues 11–12, 2010, Pages 2491–2505.
20. T. Landenfeld, A. Sadiki, J. Janicka, A Turbulence-Chemistry Interaction Model Based on a Multivariate Presumed Beta-PDF Method for Turbulent Flames, Flow, Turbulence and Combustion, Volume 68,Issue 2, 2002, pp 111–135.
21. T. Lehnhäuser and M. Schäfer. Improved linear interpolation practice for finitevolume schemes on complex grids. Int. J. Numer. Meth. Fluids, 38(7), 2002, pp 625–645.
22. N. M. Marinov: A detailed chemical kinetic model for high temperature ethanol oxidation. Int. J. Chem. Kinet., 31: 183–220, 1999.
23. A. R. Masri, J. D. Gounder, Turbulent Spray Flames of Acetone and Ethanol Approaching Extinction, Journal: Combustion Science and Technology, vol. 182, 2010, pp. 702–715.
24. A. Milford, C.B. Devaud, Investigation of an inhomogeneous turbulent mixing model for conditional moment closure applied to autoignition, Comb. and Flame, Vol. 157, Issue 8, 2010, Pages 1467–1483.
25. R. S. Miller, K. Harstad, and J. Bellan. Evaluation of eqnilibrium and non-eqnilibrium evaporation models for many gas-liquid flow simulations. Int. J. Multiphase Flow, 24:1026–1055, 1998.
26. M. Mortensen, R. W. Bilger "Derivation of the conditional moment closure eqnations for spray combustion" Combustion and Flame, Volume 156, Issue 1, 2009, Pages 62–72.
27. S. Navarro-Martinez, A. Kronenburg, LES–CMC simulations of a lifted methane flame, Proceedings of the Combustion Institute, Volume 32, Issue 1, 2009, Pages 1509–1516.
28. A. Neophytou, E. Mastorakos, R.S. Cant, Complex chemistry simulations of spark ignition in turbulent sprays, Proceedings of the Combustion Institute, Volume 33, Issue 2, 2011, Pages 2135–2142.
29. C. Olbricht, A. Ketelheun, F. Hahn and J. Janicka, Assessing the predictive capabilities of Combustion LES as applied to the Sydney flame series, Flow Turbulence and Combustion, 83 (3), 2011, pp 513–547.
30. N. Patel and S. Menon, Simulation of spray–turbulence–flame interactions in a lean direct injection combustor, Combustion and Flame, Volume 153, Issues 1–2, April 2008, Pages 228–257.
31. C. Pera, J. Réveillon, L. Vervisch, P. Domingo, Modeling subgrid scale mixture fraction variance in LES of evaporating spray, Combustion and Flame, Volume 146, Issue 4, September 2006, Pages 635–648.
32. Cécile Pera, Julien Réveillon, Luc Vervisch, Pascale Domingo, Modeling subgrid scale mixture fraction variance in LES of evaporating spray, Combustion and Flame, Volume 146, Issue 4, 2006, Pages 635–648.
33. N. Peters, Laminar diffusion flamelet models in non-premixed turbulent combustion, Progress in Energy and Combustion Science, vol. 10, Issue 3, 1984, pp 319–339.

34. S. Pichon, G. Black, N. Chaumeix, M. Yahyaoui, J.M. Simmie, H. J. Curran, R. Donohue, The combustion chemistry of a fuel tracer: Measured flame speeds and ignition delays and a detailed chemical kinetic model for the oxidation of acetone, Com. & Flame, Vol. 156, 2009, pp 494–504.
35. C. Pierce and P. Moin, Progress variable approach for large eddy simulation of turbulent non-premixed combustion, J. Fluid Mechanic, Vol. 504:73–97.
36. H.Pitsch: Large-Eddy Simulation of Turbulent Combustion, Annual Re-view of Fluid Mechanics, Vol. 38: 453–482, 2006.
37. T. Poinsot and D. Veynante, Theoretical and Numerical Combustion, 3rd Edition, (2011).
38. J. Pozorski, S.A. Apte: Filtered particle tracking in isotropic turbulence and stochastic modeling of subgrid-scale dispersion, Int. J.of Multiph. Flow, Vol. 35 (2) pp. 118–128, 2009.
39. A. Sadiki, W. Ahmadi, M. Chrigui, Toward the Impact of Fuel Evaporation-Combustion Interaction on Spray Combustion in Gas Turbine Combustion Chambers. Part I: Effect of Partial Fuel Vaporization on Spray Combustion. Chapter in Experiments and Numerical Simulations of Diluted Spray Turbulent Combustion, Proceedings of the 1st International Workshop on Turbulent Spray Combustion, Series: ERCOFTAC Series, Vol. 17, Merci, Bart; Roekaerts, Dirk; SADIKI, AMSINI (Eds.), 2011.
40. A. Sadiki, M. Chrigui, and A. Dreizler, Thermodynamically Consistent Modelling of Gas Turbine Combustion Sprays, Fluid Mechanics and Its Applications 102, Flow and Combustion in Advanced Gas Turbine Combustors DOI 10.1007/978-94-007-5320-4 3.
41. A. Sadiki, M. Chrigui, J. Janicka, M.R. Maneshkarimi, Modeling and Simulation of Effects of Turbulence on Vaporization, Mixing and Combustion of Liquid-Fuel Sprays, In: Flow Turb. Comb., 75 (1–4), (2005).
42. P. Sagaut, Large Eddy Simulation for incompressible Flows, Springer, Berlin, 2001.
43. M. Sanjosé, J.M. Senoner, F. Jaegle, B. Cuenot, S. Moreau, T. Poinsot, Fuel injection model for Euler–Euler and Euler–Lagrange large-eddy simulations of an evaporating spray inside an aeronautical combustor, International Journal of Multiphase Flow, Volume 37, Issue 5, June 2011, Pages 514–529.
44. S.S. Sazhin.: Advanced models for fuel droplet heating and evaporation, Progress in Energy and Combustion Science 32, pp. 162–214, 2006.
45. P. Schroll, A.P. Wandel, R.S. Cant, E. Mastorakos, Direct numerical simulations of autoignition in turbulent two-phase flows, Proceedings of the Combustion Institute, Volume 32, Issue 2, 2009, Pages 2275–2282.
46. W. A. Sirignano, Fluid dynamics of sprays, J. Fluids Engng. Vol. 115, 1993, pp. 345–378.
47. S. H. Stårner, J. Gounder, and A. R. Masri, Effects of turbulence and carrier fluid on simple, turbulent spray jet flames, Com. & Flame, Vol. 143, 2005, pp 420–432.
48. J.A. Van Oijen, L.P.H. De Goey, A numerical study of confined triple flames using a flamelet-generated manifold, Combust. Theory Modelling, 2004, pp 141–163.
49. O. Vermorel, S. Richard, O. Colin, C. Angelberger, A. Benkenida, D. Veynante, Towards the understanding of cyclic variability in a spark ignited engine using multi-cycle LES, Combustion and Flame, Volume 156, Issue 8, August 2009, Pages 1525–1541.
50. A. W. Vreman, B. A. Albrecht, J. A. van Oijen, L. P. H. de Goey, R. J. M. Bastiaans: Premixed and non-premixed generated manifolds in large-eddy simulation of Sandia flame D and F. Combust Flame 153, 394–416, 2008.
51. B. Wegner, A. Maltsev, C., Schneider, A., Sadiki, A., Dreizler, J., Janicka, Assessment of unsteady RANS in predicting swirl flow instability based on LES and Experiments. International Journal of Heat and Fluid Flow, 2004, 25:528–536.
52. M.R.G. Zoby, S. Navarro-Martinez, A. Kronenburg, A.J. Marquis, Evaporation rates of droplet arrays in turbulent reacting flows, Proceedings of the Combustion Institute, Volume 33, Issue 2, 2011, Pages 2117–2125.

Probability Density Function Modeling of Turbulent Spray Combustion

Rana M. Humza, Yong Hu and Eva Gutheil

Abstract Spray processes play a crucial role in liquid fueled combustion devices such as Diesel or fueled rocket engines and industrial furnaces. The combustion occurs under turbulent conditions, and a wide dynamic range of length and time scales characterize these processes, where the scales of the flow field and chemical reactions typically differ considerably. Moreover, a strong interdependence of liquid breakup and atomization, turbulent dispersion, droplet evaporation, and fuel-air mixing makes the spray modeling a challenging task. In the present chapter, a one-point one-time Eulerian statistical description of a joint mixture fraction—enthalpy probability density function (pdf) model for the gas phase is derived and modeled. A Lagrangian Monte Carlo method is used to solve the high-dimensional joint pdf transport equation. Two different mixing models, the interaction-by-exchange-with-the-mean and an extended modified Curl model, are employed in order to evaluate molecular mixing in the context of two-phase reacting flows. Moreover, a modified β function for application in turbulent spray flames, which has been proposed in an earlier study of non-reacting spray flows, is discussed in comparison with the standard β function and the transported pdf method. The modified β function is defined through two additional parameters compared to the standard form, and the choice of these parameters is discussed in the present study. A steady, two-dimensional, axisymmetric, turbulent liquid fuel/air spray flame is investigated, where both methanol and ethanol are studied. The numerical results are compared and discussed in context with the experimental data.

E. Gutheil (✉) · Y. Hu · R. M. Humza
Interdisciplinary Center for Scientific Computing, University of Heidelberg,
Im Neuenheimer Feld 368, 69120 Heidelberg, Germany
e-mail: gutheil@iwr.uni-heidelberg.de

R. M. Humza
e-mail: rana.humza@iwr.uni-heidelberg.de

Y. Hu
e-mail: yong.hu@iwr.uni-heidelberg.de

B. Merci, E. Gutheil (eds.), *Experiments and Numerical Simulations of Turbulent Combustion of Diluted Sprays,* ERCOFTAC Series 19, DOI 10.1007/978-3-319-04678-5_6,
© Springer International Publishing Switzerland 2014

1 Introduction

Spray combustion is of great practical relevance in many technical applications such as Diesel engines, rocket engines, and industrial furnaces. In these combustion devices, the liquid fuel is injected through suitable nozzles into the combustion chamber, and it is atomized to form a dilute turbulent spray.

For the chemically reacting spray flows, the detailed combustion chemistry and its intense interaction with hydrodynamic turbulence and the evaporation process still pose a challenging modeling task, and a variety of different solutions and closure schemes are proposed and developed. For gas combustion, the flamelet model [39], conditional moment closure (CMC) [35] and transported probability density function (pdf) [42] methods are the most successful. Flamelet models, based on pre-calculated chemistry state-relationship tables and assuming that the thin laminar flamelet forms the local flame structure in a turbulent flame, are originally proposed for gaseous flames [39] and later extended to gas-liquid two-phase reacting flows [24]. In this modeling approach, the detailed, multidimensional chemistry is mapped to a few characteristic, controlling variables. In the first order formulation, the CMC model [35] makes use of conditional averages of reactive scalar values to close the chemical reaction terms in the governing equations with the assumption that the turbulent fluctuations of the gas phase composition conditioned on the mixture fraction can be neglected. Recently, Mortensen and Bilger [35] derived the CMC equations for a general two-phase reacting flow with additional terms arising from the effect of transport at the inter-phase surfaces, and De et al. [8] used this extension of CMC to simulate the non-reacting and reacting turbulent spray flows combined with a two equation turbulence model. The present simulations employ the spray flamelet model [24] for turbulent spray combustion.

Concerning large eddy simulations (LES) of the turbulent flow field, the filtering implies that the scales below the filter width are not resolved and must be modeled [33]. However, in turbulent spray combustion, molecular mixing and chemical reactions occur at unresolved scales. Hence, modeling the processes in the sub-grid range imposes additional challenges. In Reynolds averaged Navier-Sokes (RANS) models, where most often the (extended) k-ϵ model is applied, the assumption of isotropy of turbulence and the inability of a proper description of counter-gradient diffusion cause problems. Moreover, transported (joint) pdf models are also considered. They are based on using the concept of a one–point one–time Eulerian joint pdf for a set of variables that describe local velocity and/or composition fields. By solving the modeled transport equations for the joint pdf, a complete description of the reacting flow with increased statistical information is achieved. Here, all moments of the dependent quantities can be evaluated from the resulting joint statistic pdf. For pure gas combustion, Lundgren [28] derived and modeled a transport equation for the joint pdf of velocity components, and Dopazo and O'Brien [10] solved a modeled transported equation for the joint pdf of chemical composition and gas temperature. Then the velocity-composition joint pdf transport model was shown to be good basis for modeling turbulent reacting flows with exact treatment of turbulent convection and chemical reactions [41]. The pdf method received more attention since Pope's work [40, 42], and a Lagrangian Monte Carlo algorithm was proposed to solve the transport equation of the joint pdf. With this advancement in numerical algorithms, the pdf method has achieved success in the prediction of local extinction and re-ignition [47], and the formation of pollutants such as nitric oxide and soot [32] in single gas phase reacting flows [19]. The major advantages of the pdf method are the exact and closed form of convection, body force, mean pressure gradient and finite-rate chemistry in the modeled transport equation. Moreover, the Lagrangian pdf takes the long memory of the hydrodynamic turbulence fully into account. The past history of all the fluid particles' properties can be completely described by the multi-time Lagrangian pdf.

The transported pdf method was applied to simulate multiphase flows since the 1990s. Raju [43] and Durand et al. [11] described the gas phase in turbulent two-phase flows with a pdf transport equation. Rumberg and Rogg [45], Zhu et al. [48], and Pai and Subramaniam [38] deduced a unified joint pdf transport equation of all liquid-phase and gas-phase dependent variables. Garg et al. [14] developed accurate numerical methods to solve these pdf equations. Liu et al. [27] and Naud [36] dealt with the joint pdf transport equation of the properties of droplets and gas eddies seen by droplets. Ge and Gutheil [15] deduced and solved the transport equation of the pdf of mixture fraction in turbulent spray flows, and they suggested a modified presumed pdf for turbulent mixing in sprays. Ge and Gutheil [16] and Ge et al. [17, 18] further developed a joint mixture fraction-enthalpy pdf and a joint velocity-mixture fraction pdf to simulate the turbulent reacting and non-reacting spray flows, respectively. Kung and Haworth [26] applied a transported pdf equation to engine simulation. Anand and Jenny [2] used a joint gas phase velocity-composition-frequency pdf for an evaporating spray, where the droplet-seen gas velocity was included in the droplet density function. Heye et al. [22] applied a joint mixture fraction and progress variable to large eddy simulation of a turbulent ethanol spray flame, and a consistent Lagrangian Monte-Carlo algorithm for solving the modeled pdf transport equation was developed.

The present study concerns turbulent spray flame modeling, where both a transported joint pdf equation of the mixture fraction and enthalpy as well as presumed pdf modeling are considered. In the latter situation, the four parameter modified β function [15] that has been suggested for use in non-reacting spray flows and which has been used in reacting spray flows [29] is re-investigated for use in turbulent spray flames. Both methanol and ethanol spray flames in air are studied in the framework of a spray flamelet model accounting for spray evaporation effects in turbulent combustion.

After the presentation of the mathematical model, results for both flames are presented and discussed.

2 Mathematical Model

The mathematical modeling of a turbulent spray flame concerns a two-phase problem, where a dilute spray is carried by a continuous gas phase. The gas phase conservation equations include source terms to account for the interaction of the gas with the spray. The system of the Favre-averaged governing equations is closed using an extended $k-\epsilon$ model [23], which accounts for the spray evaporation through additional source terms. The present simulations concern both a turbulent methanol/air flame [31] and an ethanol/air spray flame [12, 17], and the detailed chemistry is included using laminar spray flamelet libraries for methanol/air [21] and ethanol/air [20]. The effect of turbulence on chemical reactions is modeled using probability density functions to account for statistical fluctuations of characteristic variables of the flow field. The pdf can be either presumed or transported, and both formulations are analyzed and discussed in the following sections.

2.1 *Governing Equations*

The governing equations consist of gas phase conservation equations with appropriate spray source terms. These equations are closed by the mathematical description of the liquid phase.

2.1.1 Gas Phase Equations

The mathematical formulation for the gas phase uses Favre-averaged variables in order to account for variable density of the chemically reactive spray flow at low Mach number. The Favre-averaged conservation equations of mass, momentum and energy assuming no swirl and low Mach number for a dilute spray yield

$$\frac{\partial \overline{\rho}}{\partial t} + \frac{\partial (\overline{\rho}\,\widetilde{u}_j)}{\partial x_j} = \overline{S}_v \tag{1}$$

$$\begin{aligned}\frac{\partial (\overline{\rho}\,\widetilde{u}_i)}{\partial t} + \frac{\partial (\overline{\rho}\,\widetilde{u}_j \widetilde{u}_i)}{\partial x_j} - \frac{\partial}{\partial x_j}\left(\Gamma_{\tilde{u}_i,\text{eff}} \frac{\partial \widetilde{u}_i}{\partial x_j}\right) &= -\frac{\partial \overline{p}}{\partial x_i} + \frac{\partial}{\partial x_j}\left(\Gamma_{\tilde{u}_i,\text{eff}} \frac{\partial \widetilde{u}_j}{\partial x_i}\right) \\ &\quad - \frac{\partial}{\partial x_i}\left(\frac{2}{3}\Gamma_{\tilde{u}_i,\text{eff}} \frac{\partial \widetilde{u}_k}{\partial x_k} + \frac{2}{3}\rho \tilde{k}\right) + \overline{\rho} g_{i+} \overline{S}_{m,i}\end{aligned} \tag{2}$$

$$\frac{\partial (\overline{\rho}\tilde{h})}{\partial t} + \frac{\partial (\overline{\rho}\,\widetilde{u}_j \tilde{h})}{\partial x_j} - \frac{\partial}{\partial x_j}\left(\Gamma_{\tilde{h},\text{eff}} \frac{\partial \tilde{h}}{\partial x_j}\right) = \frac{\partial \overline{p}}{\partial t} + \overline{S}_e. \tag{3}$$

The mixture fraction is defined following Chap. 4 of this book, and it is based on the chemical element carbon, C, which is omitted in the following notation for simplicity of the formulation. The transport equation for the mixture fraction may be derived as outlined in Chap. 4 [37], and it is used here in connection with the presumed pdf studies. Moreover, the formulation of the mixture fraction is necessary to include detailed chemical reactions through use of the spray flamelet model [17, 16, 24]

$$\frac{\partial(\overline{\rho}\widetilde{\xi})}{\partial t} + \frac{\partial(\overline{\rho}\,\widetilde{u}_j\widetilde{\xi})}{\partial x_j} - \frac{\partial}{\partial x_j}\left(\Gamma_{\tilde{\xi},\mathrm{eff}}\frac{\partial\widetilde{\xi}}{\partial x_j}\right) = \overline{S}_v. \tag{4}$$

The transport equation for the variance of the mixture fraction, $\widetilde{\xi''^2}$, may be derived following earlier work [23]. New terms associated with the spray source term, S_v, are modeled [23], so that the closed transport equation yields

$$\frac{\partial(\overline{\rho}\widetilde{\xi''^2})}{\partial t} + \frac{\partial(\overline{\rho}\,\widetilde{u}_j\widetilde{\xi''^2})}{\partial x_j} - \frac{\partial}{\partial x_j}\left(\Gamma_{\widetilde{\xi''^2},\mathrm{eff}}\frac{\partial\widetilde{\xi''^2}}{\partial x_j}\right)$$
$$= 2\Gamma_{\widetilde{\xi''^2},\mathrm{eff}}\left(\frac{\partial\widetilde{\xi}}{\partial x_k}\right)^2 - 2\overline{\rho}\frac{\widetilde{\epsilon}}{\widetilde{k}}\widetilde{\xi''^2} + \overline{S}_v\widetilde{\xi''^2}(1-2\widetilde{\xi})/\widetilde{\xi}. \tag{5}$$

For the transported pdf of the mixture fraction and the enthalpy of the gas phase, the Eqs. 3, 4 and 5 are not needed.

In the above equations, $\overline{\rho}$ is the density of the gas mixture and $\overline{p}$ denotes pressure, where the bar denotes time-averaged values. S_v, $\vec{S}_m$ and S_e are the source terms for the mass, momentum and energy [15] due to spray evaporation. $\Gamma_{\tilde{\Phi},\mathrm{eff}} = \eta_l/\sigma_{\Phi,l} + \eta_t/\sigma_{\Phi,t}$ denotes the effective exchange coefficient for variable Φ, η_l is the molecular viscosity, $\sigma_{\Phi,l}$ and $\sigma_{\Phi,t}$ are the laminar and turbulent Prandtl-Schmidt numbers, respectively, and η_t the turbulent viscosity given by $\eta_t = C_\eta\overline{\rho}\widetilde{k}^2/\widetilde{\epsilon}$ with $C_\eta = 0.09$. Here $\widetilde{k}$ and $\widetilde{\epsilon}$ denote the turbulent kinetic energy and its dissipation rate, and they are obtained by solving an extended $k-\epsilon$ model [23]. The transport equations for $\widetilde{k}$ and $\widetilde{\epsilon}$ are

$$\frac{\partial(\overline{\rho}\widetilde{k})}{\partial t} + \frac{\partial(\overline{\rho}\,\widetilde{u}_j\widetilde{k})}{\partial x_j} - \frac{\partial}{\partial x_j}\left(\Gamma_{\tilde{k},\mathrm{eff}}\frac{\partial\widetilde{k}}{\partial x_j}\right) = G_k - \overline{\rho}\widetilde{\epsilon} - \frac{2}{3}\overline{\rho}\widetilde{k}\frac{\partial\widetilde{u}_j}{\partial x_j} + \overline{S}_k, \tag{6}$$

$$\frac{\partial(\overline{\rho}\,\widetilde{\epsilon})}{\partial t} + \frac{\partial(\overline{\rho}\,\widetilde{u}_j\widetilde{\epsilon})}{\partial x_j} - \frac{\partial}{\partial x_j}\left(\Gamma_{\tilde{\epsilon},\mathrm{eff}}\frac{\partial\widetilde{\epsilon}}{\partial x_j}\right) = (C_{\epsilon 1}G_k - C_{\epsilon 2}\overline{\rho}\,\widetilde{\epsilon})\frac{\widetilde{\epsilon}}{\widetilde{k}}$$
$$-\left(\frac{2}{3}C_{\epsilon 1} - C_{\epsilon 3}\right)\overline{\rho}\,\widetilde{\epsilon}\frac{\partial\widetilde{u}_j}{\partial x_j} + \overline{S}_\epsilon, \tag{7}$$

with

$$G_k = \eta_t\left\{\left(\frac{\partial\widetilde{u}_j}{\partial x_i} + \frac{\partial\widetilde{u}_i}{\partial x_j}\right) - \frac{2}{3}\left(\frac{\partial\widetilde{u}_k}{\partial x_k}\right)\delta_{ij}\right\}\frac{\partial\widetilde{u}_i}{\partial x_j}. \tag{8}$$

Here $\overline{S}_k$ is the spray source term for the turbulent kinetic energy, $\widetilde{k}$, and $\overline{S}_\epsilon = c_S \frac{\widetilde{\epsilon}}{\widetilde{k}} \overline{S}_k$. The model constants are $C_{\epsilon 1} = 1.40$, $C_{\epsilon 2} = 1.90$, $C_{\epsilon 3} = -1.0$, and $c_S = 1.5$, which are standard parameters.

2.1.2 Liquid Phase Equations

The Abramzon and Sirignano model [1] is used to calculate the evaporated droplet mass, $\dot{m}$, in a convective flow field

$$\dot{m} = 2\pi \rho_f D_f r \,\widetilde{\mathrm{Sh}} \ln(1 + B_M), \tag{9}$$

where the subscript f denotes properties in the gas film, and it is determined according to the 1/3 rule [25] while r denotes the droplet radius. $\widetilde{\mathrm{Sh}}$ is the modified Sherwood number and B_M is the Spalding mass transfer number given by $B_M = (Y_{Fs} - Y_{F\infty})/(1 - Y_{Fs})$, where Y_{Fs} and $Y_{F\infty}$ are the mass fractions of fuel vapor at the droplet surface and in the surrounding gas respectively, and, following Dalton's law [3], the fuel vapor mass fraction at the droplet surface, Y_{Fs}, is given by $Y_{Fs} = M_F/[M_F + M(p/p_F - 1)]$. The quantities M_F and p_F denote molar mass and vapor pressure of fuel, while M and p represent the mean molar mass of the surrounding gas and atmospheric pressure, respectively. Vapor pressure of fuel is calculated using the Clausius–Clapeyron equation.

Assuming the droplet velocity to evolve by interactive drag applied by the surrounding gas and gravity, the following relation is used to express the equation of droplet motion [4]

$$\frac{\mathrm{d}\mathbf{v}_p}{\mathrm{d}t} = \frac{3}{8}\frac{1}{r}\frac{\overline{\rho}}{\rho_l} C_D (\mathbf{u} - \mathbf{v}_p)|\mathbf{u} - \mathbf{v}_p| + \vec{g}, \tag{10}$$

where $\overline{\rho}$ and $\mathbf{u}$ are the density and velocity of the surrounding gas, while ρ_l, C_D, and $\mathbf{g}$ are liquid density, drag coefficient, and gravitational acceleration, respectively. The drag coefficient C_D is calculated as a function of the droplet Reynolds number, $\mathrm{Re}_d = 2r\overline{\rho}|\mathbf{u} - \mathbf{v}_p|/\eta_f$, where η_f is the mean dynamic viscosity in the film, as [46]

$$C_D = \begin{cases} \frac{24}{\mathrm{Re}_d}\left(1 + 0.15\,\mathrm{Re}_d^{0.687}\right) & \text{if } \mathrm{Re}_d < 10^3 \\ 0.44 & \text{if } \mathrm{Re}_d \geq 10^3 \end{cases}. \tag{11}$$

For the evolution of droplet temperature, an infinite thermal conductivity model is used [1], which is a good assumption for the present small droplet with high volatility of the fuels under consideration at atmospheric pressure. Thus, the liquid temperature can be expressed as

$$mC_{p,l}\frac{\mathrm{d}T_s}{\mathrm{d}t} = \dot{m}\left[\frac{C_{pF}\left(T_g - T_s\right)}{B_T} - L_V(T_s)\right], \tag{12}$$

where $m = 4/3\pi\rho_l r^3$ denotes the droplet mass, $C_{p,l}$ and C_{pF} are the specific heat capacities of the liquid at constant pressure and of the gas in the film, respectively.

Table 1 Spray source terms

S_v	$\mathbf{S}_m$	S_e	S_k	S_ϵ
$\sum_{k=1}^{n_p} N_{p,k}\dot{m}$	$\sum_{k=1}^{n_p} N_{p,k}\left[-m\frac{\mathrm{d}\vec{v}_p}{\mathrm{d}t}+\dot{m}\vec{v}_p\right]$	$\sum_{k=1}^{n_p} N_{p,k}\dot{m}h_d$	$\sum_{k=1}^{n_p} N_{p,k}\dot{m}\mathbf{u}''\mathbf{v}_p$	$c_S\sum_{k=1}^{n_p} N_{p,k}\dot{m}\mathbf{u}''\mathbf{v}_p\frac{\tilde{\epsilon}}{k}$

T_s is the uniform droplet temperature, T_g denotes the gas temperature and L_V is the temperature dependent latent heat of vaporization. $B_T = (1 + B_M)^\phi - 1$ is the Spalding heat transfer number, which is calculated in terms of the mass transfer number [1], where the exponent is $\phi = (C_{pF}/C_{pg})(\widetilde{\mathrm{Sh}}/\widetilde{\mathrm{Nu}})(1/\mathrm{Le})$ [1]. Here, C_{pg} is the specific heat capacity of the gas and $\widetilde{\mathrm{Nu}}$ is the modified Nusselt number [1].

The liquid phase equations are solved using a stochastic parcel method. The source terms due to spray evaporation S_v, $\mathbf{S}_m$, S_e, S_k, and S_ϵ for the gas phase conservation Eqs. 1–7 are computed using the particle-source-in-cell (PSIC) method [6, 23, 24, 17], where the droplets are considered as point sources. To determine the droplet source in a control cell, the droplet position is determined at first using the relation

$$\mathbf{x}_p^{n+1} = \mathbf{x}_p^n + \mathbf{v}_p^n \Delta t. \tag{13}$$

The mean value of a variable Φ in control volume V around the node (i, j) is defined as

$$\overline{\Phi}|_V = \frac{\sum_{k=1}^{n_p} N_{p,k}\Phi|_{i,j}}{\sum_{k=1}^{n_p} N_{p,k}}, \tag{14}$$

where $N_{p,k}$ denotes the number of droplets in k^{th} parcel and n_p is the total number of parcels. The spray source terms for a control volume V can be written as shown in Table 1.

The gas and liquid phase equations are strongly coupled, and a hybrid numerical solution scheme is used to solve the equations, for details see [23].

This study concerns both presumed and transported pdf methods. In case of the latter choice, the transport equation of the pdf of one or more than one variables must be derived and solved numerically. Previous work concerned a univariate transported pdf equation for the mixture fraction [15], the joint pdf of gas velocity and mixture fraction in non-reacting spray flows [18], and the mixture fraction and gas enthalpy [17], where the last formulation neglects the source term of the mass, c.f. Eq. 1, in the derivation of the joint pdf equation when the continuity equation is used to simplify equations, which is valid for highly dilute sprays. This assumption is relaxed in the derivation of the following general joint bivariate pdf formulation of the mixture fraction and gas enthalpy.

2.2 *Joint Probability Density Function of Mixture Fraction and Gas Enthalpy*

This section presents the derivation of the joint transported pdf of the mixture fraction and the gas enthalpy. The derivation follows previous work [17, 18]. A fine-grained, one-point one-time Eulerian, joint mixture fraction and enthalpy probability density function $f^*(\zeta,\eta;\mathbf{x},t)$ is defined for the gas phase of turbulent spray flames

$$f^*(\zeta,\eta;\mathbf{x},t)=\delta(\xi(\mathbf{x},t)-\zeta)\delta(h(\mathbf{x},t)-\eta). \quad (15)$$

Here, ξ and h are mixture fraction and gas enthalpy in physical space, and ζ and η the corresponding values in sample space. Ensemble averaging of this fine-grained probability density function yields

$$f(\zeta,\eta;\mathbf{x},t)=\langle f^*(\zeta,\eta;\mathbf{x},t)\rangle=\langle\delta(\xi(\mathbf{x},t)-\zeta)\delta(h(\mathbf{x},t)-\eta)\rangle. \quad (16)$$

The conditional mean of any function $Q=Q(\mathbf{x},t)$ can be related to $f(\zeta,\eta;\mathbf{x},t)$ by

$$\langle Q(\mathbf{x},t)f^*(\zeta,\eta;\mathbf{x},t)\rangle=\langle Q(\mathbf{x},t)|\zeta,\eta\rangle f(\zeta,\eta;\mathbf{x},t). \quad (17)$$

Considering the properties of the Dirac delta function, the material derivative of the fine-grained pdf may be written as

$$0=\frac{Df^*}{Dt}=\frac{\partial f^*}{\partial t}+\frac{\partial f^*}{\partial \mathbf{x}}\frac{d\mathbf{x}}{dt}+\frac{\partial f^*}{\partial \zeta}\frac{d\zeta}{dt}+\frac{\partial f^*}{\partial \eta}\frac{d\eta}{dt}. \quad (18)$$

With the sifting property of Dirac delta function, Eq. 18 may be re-written as

$$\begin{aligned}\rho\frac{\partial f^*}{\partial t}+\rho u_j\frac{\partial f^*}{\partial x_j}&=-\rho\frac{\partial f^*}{\partial \zeta}\frac{d\zeta}{dt}-\rho\frac{\partial f^*}{\partial \eta}\frac{d\eta}{dt}\\&=-\frac{\partial}{\partial\zeta}(\rho\frac{d\xi}{dt}f^*)-\frac{\partial}{\partial\eta}(\rho\frac{dh}{dt}f^*).\end{aligned} \quad (19)$$

Use of Eqs. 16 and 17 leads to

$$\rho\frac{\partial f}{\partial t}+\rho u_j\frac{\partial f}{\partial x_j}=-\frac{\partial}{\partial\zeta}(\rho\langle\frac{d\xi}{dt}|\zeta,\eta\rangle f)-\frac{\partial}{\partial\eta}(\rho\langle\frac{dh}{dt}|\zeta,\eta\rangle f). \quad (20)$$

Substitution of the instantaneous continuity equation $\partial\rho/\partial t+\partial(\rho u_j)/\partial x_j=S_v$ into the above equation and considering the joint mass density function $F(\zeta,\eta;\mathbf{x},t)=\rho f(\zeta,\eta;\mathbf{x},t)$, yields

$$\frac{\partial F}{\partial t}+\frac{\partial(u_jF)}{\partial x_j}-\langle\frac{S_v}{\rho}|\zeta,\eta\rangle F=-\frac{\partial}{\partial\zeta}(\langle\frac{d\xi}{dt}|\zeta,\eta\rangle F)-\frac{\partial}{\partial\eta}(\langle\frac{dh}{dt}|\zeta,\eta\rangle F). \quad (21)$$

In this transport equation of mass density function F, the terms on the left hand side are closed including the mass evaporation source, whereas the terms on the right hand side are unclosed. Based on the instantaneous conservation equations for the gas enthalpy and mixture fraction, the unclosed terms may be written as

$$\begin{aligned}-\frac{\partial}{\partial\zeta}(\langle\frac{d\xi}{dt}|\zeta,\eta\rangle F)=&-\frac{1}{\rho}\langle(1-\xi)S_v\rangle\frac{\partial F}{\partial\zeta}-\frac{\partial}{\partial\zeta}(\frac{1}{\rho}\langle\frac{\partial}{\partial x_j}(\rho D_M\frac{\partial\xi}{\partial x_j})\\&+S_v^{'}+\langle\xi S_v\rangle-\xi S_v|\zeta,\eta\rangle F)\end{aligned} \quad (22)$$

$$-\frac{\partial}{\partial \eta}(\langle \frac{dh}{dt}|\zeta,\eta\rangle F) = -\frac{1}{\rho}\langle S_e - hS_v\rangle \frac{\partial F}{\partial \eta} - \frac{\partial}{\partial \eta}(\frac{1}{\rho}\langle \frac{\partial}{\partial x_j}(\rho D_h \frac{\partial h}{\partial x_j}) + S_e^{'} + \langle hS_v\rangle - hS_v|\zeta,\eta\rangle F). \tag{23}$$

Insertion into Eq. 21 yields

$$\begin{aligned}\frac{\partial F}{\partial t} + \frac{\partial (u_j F)}{\partial x_j} - \frac{\langle S_v\rangle}{\rho}F + \frac{1}{\rho}\langle (1-\xi)S_v\rangle \frac{\partial F}{\partial \zeta} + \frac{1}{\rho}\langle S_e - hS_v\rangle \frac{\partial F}{\partial \eta} \\ = -\frac{\partial}{\partial \zeta}(\frac{1}{\rho}\langle \frac{\partial}{\partial x_j}(\rho D_M \frac{\partial \xi}{\partial x_j}) + S_\xi^{(f)}|\zeta,\eta\rangle F) \\ - \frac{\partial}{\partial \eta}(\frac{1}{\rho}\langle \frac{\partial}{\partial x_j}(\rho D_h \frac{\partial h}{\partial x_j}) + S_h^{(f)}|\zeta,\eta\rangle F),\end{aligned} \tag{24}$$

where the fluctuations of spray source terms are

$$S_\xi^{(f)} = (1-\xi^{'})S_v^{'} \tag{25}$$

$$S_h^{(f)} = (1-h^{'})S_e^{'}. \tag{26}$$

In Eq. 24 the terms on the right hand side appear in unclosed form. They account for the effects of the molecular diffusion and the fluctuations in spray source terms, respectively. In the present work, the fluctuations of spray source terms, i.e. Eqs. 25 and 26, are considered to be very small compared to the mean values, and they are neglected. The molecular diffusion and conduction, respectively, are approximated through use of an extended interaction-by-exchange-with-the-mean (IEM) model

$$\frac{d\xi^*(t)}{dt} = -\frac{1}{2}\frac{\widetilde{\epsilon}}{\widetilde{k}}C_\phi[\xi^*(t) - \widetilde{\xi}] + [1-\widetilde{\xi}]\frac{\langle S_v\rangle}{\langle \rho\rangle} \tag{27}$$

$$\frac{dh^*(t)}{dt} = -\frac{1}{2}\frac{\widetilde{\epsilon}}{\widetilde{k}}C_\phi[h^*(t) - \widetilde{h}] + \frac{\langle S_e\rangle - \widetilde{h}\langle S_v\rangle}{\langle \rho\rangle} \tag{28}$$

where C_ϕ is the mixing model constant. Compared to an earlier formulation [15] for the joint pdf of mixture fraction and velocity, the present formulation includes a new term (the second part of the last term in Eq. 27), which appears through use of the continuity equation, Eq. 1, which was neglected in the earlier formulation. A revised formulation [18] for the joint mixture fraction–velocity joint pdf accounts for this. Similarly, a corresponding term appears in Eq. 28. The effect of these new terms will be discussed in Sect. 3.

Cao et al. [5] investigated the validity and efficiency of mixing models on gas flames. They found that the IEM should be replaced by the Euclidean minimal spanning tree (EMST) or modified Curl (MC) mixing model for jet flames. Therefore, an extended modified Curl (MC) is also employed in order to evaluate the effects of micro-mixing. The MC model [13] is based upon Curl's particle interaction model [7]. Two equal-weight stochastic particles, denoted by p_1 and p_2, are selected at random from the ensemble and after mixing, their mixture fractions and enthalpies are given by

$$\xi^{(p_1,new)} = \xi^{(p_1)} + \frac{1}{2}a\left(\xi^{(p_2)} - \xi^{(p_1)}\right) \tag{29}$$

$$\xi^{(p_2,new)} = \xi^{(p_2)} + \frac{1}{2}a\left(\xi^{(p_1)} - \xi^{(p_2)}\right) \tag{30}$$

$$h_s^{(p_1,new)} = h_s^{(p_1)} + \frac{1}{2}a\left(h_s^{(p_2)} - h_s^{(p_1)}\right) \tag{31}$$

$$h_s^{(p_2,new)} = h_s^{(p_2)} + \frac{1}{2}a\left(h_s^{(p_1)} - h_s^{(p_2)}\right). \tag{32}$$

The coefficient a is a random number between zero and unity.

Solution strategies for the joint pdf transport equation are presented and discussed in previous work [15–18].

2.3 Presumed Probability Density Function Modeling

The turbulence in combustion systems makes the application of stochastic techniques inevitable, which account for the statistical fluctuations of the characteristic variables. In the realm of stochastic processes, the presumed pdf methods [9, 11] are widely applied because of their numerical ease and computational efficiency compared to transported pdf methods.

In turbulent gas flames, detailed chemistry is modeled using a laminar flamelet library, where each of the laminar flamelets is characterized by the mixture fraction ξ and its scalar dissipation rate, $\chi = 2D\,(\partial\xi/\partial x_i)^2$, through [39]

$$\widetilde{\phi} = \int_0^\infty \int_0^1 \phi \widetilde{P}(\xi, \chi) \mathrm{d}\xi \mathrm{d}\chi, \tag{33}$$

where $\widetilde{\phi}$ is a Favre averaged scalar variable. In spray flames, the effect of evaporation must also be taken into account. Structures of laminar spray flames depend not only on mixture fraction and its scalar dissipation rate but also on initial droplet size and velocity as well as global equivalence ratio [20, 21], and a spray flamelet model [24] has been developed, which accounts for these additional parameters, leading to [24]

$$\widetilde{\phi} = \int_{-\infty}^\infty \int_0^\infty \int_0^\infty \int_0^\infty \int_0^1 \phi \widetilde{P}(\xi, \chi, E, R_0, v_0) \mathrm{d}\xi \mathrm{d}\chi \mathrm{d}E \mathrm{d}R_0 \mathrm{d}v_0. \tag{34}$$

Olguin and Gutheil [37] suggest a revised formulation of the flamelet formulation, cf. Chap. 4, but the present chapter refers to the former formulation because the pdf of the evaporation term is still under investigation. Thus, the key question is to define $\widetilde{P}(\xi, \chi, E, R_0, v_0)$. Since the initial droplet size and initial droplet velocity are taken according to physical conditions and they do not change during combustion, it is important to establish the statistical expressions of all dependent variables. Assuming statistical independence of all variables [24, 39], their joint pdf can be defined as a product of marginal pdfs

$$P(\xi, \chi, E, R_0, v_0) = P_\xi(\xi)P_\chi(\chi)P_E(E)P_{R_0}(R_0)P_{v_0}(v_0). \tag{35}$$

Thus, the statistical distributions of the dependent variables are needed. Presently, the pdf of the equivalence ratio is taken as Dirac delta function with the value of the global equivalence ratio, the library for the initial droplet radius contains the initial radii of 10 and 25 μm, and the initial droplet velocity is taken to be fixed.

The scope of the present section is the description of turbulent mixing. In gas combustion, the presumed β pdf is applied because of its boundedness between zero and unity, which matches the physical requirements of the mixture fraction, and it is flexible in shape.

2.3.1 β Distribution

In probability theory, the β distribution (also called β distribution of first kind) refers to a family of continuous probability distributions defined over the interval [0,1]. In Bayesian analysis, it serves as conjugate prior of Binomial, Bernoulli and geometric distributions. There are two parameters of the β distribution, namely a and b, which appear as exponents of a random variable and control the shape of the β pdf. The pdf of random variable S is defined as

$$P_S(s; a, b) = \frac{s^{a-1}(1-s)^{b-1}}{B(a,b)}, \tag{36}$$

where $B(a,b)$ denotes the beta function of a and b defined by

$$B(a,b) = \int_0^1 t^{a-1}(1-t)^{b-1}\mathrm{d}t = \frac{\Gamma(a)\Gamma(b)}{\Gamma(a+b)}. \tag{37}$$

The mean, E(S), and variance, Var(S), are given by

$$\mathrm{E}(S) = \frac{a}{a+b}, \tag{38}$$

$$\mathrm{Var}(S) = \frac{\mathrm{E}(S)(1-\mathrm{E}(S))}{1+a+b}. \tag{39}$$

In turbulent (spray) combustion, a simple way is to assume that the pdf of the mixture fraction, ξ, follows a β distribution [9, 11, 44]. Thus,

$$P_\xi(\xi) = \frac{\Gamma(a+b)}{\Gamma(a)\Gamma(b)}\xi^{a-1}(1-\xi)^{b-1}. \tag{40}$$

Using the local values of $\widetilde{\xi}$ and $\widetilde{\xi''^2}$, the shape parameters a and b are calculated with Eqs. 38 and 39 as

$$a = \widetilde{\xi}\left[\frac{\widetilde{\xi}(1-\widetilde{\xi})}{\widetilde{\xi''^2}} - 1\right] \tag{41}$$

$$b = (1-\widetilde{\xi})\left[\frac{\widetilde{\xi}(1-\widetilde{\xi})}{\widetilde{\xi''^2}} - 1\right]. \tag{42}$$

The parameters a and b vary throughout the turbulent flow field since they depend on the Favre mean values of the mixture fraction, $\widetilde{\xi}$, and its scalar dissipation rate, $\widetilde{\xi''^2}$, which are obtained from the solution of Eqs. 4 and 5.

For $a < 1$, the pdf has a peak value at the left flank, while for $b < 1$, the pdf has a peak at the right flank. For $a > 1$ or $b > 1$, the value of pdf is zero at the left or right flank, respectively. Thus, the possible shapes are U shape pdf ($a < 1, b < 1$), J shape pdf ($a \geq 1, b < 1$), reverse J shape pdf ($a < 1, b \geq 1$) and bell shape or Gaussian shape pdf ($a > 1, b > 1$). Furthermore, the pdf is left skewed for ($a < b$), right skewed for ($a > b$) and symmetric for ($a = b$).

2.3.2 Modified β Distribution

Considering presumed pdf methods for turbulent spray flows, it turns out [16, 34] that the presumed β function is not a good choice if spray evaporation occurs. Therefore, Ge and Gutheil [15, 16] suggested a four parameter β distribution including two additional parameters $\xi_{\min}$ and $\xi_{\max}$. The model was developed for non-reacting sprays [15], and it is applied to reacting spray flows [16]. The pdf is given by

$$P_{\xi}(\xi) = \frac{\Gamma(a+b)}{\Gamma(a)\Gamma(b)}(\xi_{\max}-\xi_{\min})^{1-a-b}(\xi-\xi_{\min})^{a-1}(\xi_{\max}-\xi)^{b-1}. \tag{43}$$

The modified pdf equation is equivalent to a revised β distributed mixture fraction variable, ξ_1

$$\xi_1 = \frac{\xi - \xi_{\min}}{\xi_{\max} - \xi_{\min}}. \tag{44}$$

The mean and variance of modified β distribution can be calculated by using the rule of change of variables, which yields

$$\widetilde{\xi} = \xi_{\min} + \frac{a}{a+b}(\xi_{\max} - \xi_{\min}) \tag{45}$$

$$\widetilde{\xi''^2} = \frac{(\tilde{\xi} - \xi_{\min})(\xi_{\max} - \tilde{\xi})}{1+a+b}. \tag{46}$$

The shape parameters are computed using Eqs. 45 and 46 through the expressions

$$a = \frac{(\tilde{\xi} - \xi_{\min})}{(\xi_{\max} - \xi_{\min})}\left[\frac{(\tilde{\xi} - \xi_{\min})(\xi_{\max} - \tilde{\xi})}{\widetilde{\xi''^2}} - 1\right] \tag{47}$$

$$b = \frac{(\xi_{\max} - \widetilde{\xi})}{(\xi_{\max} - \xi_{\min})} \left[\frac{(\tilde{\xi} - \xi_{\min})(\xi_{\max} - \widetilde{\xi})}{\widetilde{\xi''^2}} - 1 \right]. \tag{48}$$

It is important to note that the coefficient $\frac{\Gamma(a+b)}{\Gamma(a)\Gamma(b)}$ is the normalizing condition in Eq. 43, which assures that $\int_{\xi_{\min}}^{\xi_{\max}} P_{\xi}(\xi)\mathrm{d}\xi = 1$. An obvious question that arises is how to choose $\xi_{\min}$ and $\xi_{\max}$. Previous results [15] concerning a non-reacting methanol/air spray flow suggests that the values may be associated with the standard deviation of the distribution of the mixture fraction, $\widetilde{\xi}$, and the comparison of presumed pdf and transported pdf suggests the choice

$$\xi_{\min} = \widetilde{\xi} - \sqrt{\widetilde{\xi''^2}} \tag{49}$$

$$\xi_{\max} = \widetilde{\xi} + \sqrt{\widetilde{\xi''^2}}. \tag{50}$$

It can be seen that both a and b equal zero in Eqs. 47 and 48 for the above choice of $\xi_{\min}$ and $\xi_{\max}$, which yields an undefined value of the coefficient $\frac{\Gamma(a+b)}{\Gamma(a)\Gamma(b)}$ in Eq. 43. Since this coefficient is merely the normalizing condition, which in this case can be replaced by $\left[\int_{\xi_{\min}}^{\xi_{\max}} (\xi_{\max} - \xi_{\min})^{1-a-b}(\xi - \xi_{\min})^{a-1}(\xi_{\max} - \xi)^{b-1}\mathrm{d}\xi\right]^{-1}$. The formulation proposed in Eqs. 49 and 50 may further be generalized to

$$\xi_{\min} = \widetilde{\xi} - c\sqrt{\widetilde{\xi''^2}} \tag{51}$$

$$\xi_{\max} = \widetilde{\xi} + d\sqrt{\widetilde{\xi''^2}}, \tag{52}$$

where $c, d \in \mathbb{R}^+$. The choice of $c = d$ enforces $a = b$, and a symmetric shape of the modified pdf is obtained. Luo et al. [29] adapted the modified β distribution by assuming $\xi_{\min} = 0$ and $\xi_{\max} = \widetilde{\xi} + 2\sqrt{\widetilde{\xi''^2}}$, which corresponds to the choice of $d = 2$ in Eq. 52.

In the results' section, different choices of $\xi_{\min}$ and $\xi_{\max}$ will be presented and discussed.

3 Results and Discussion

In the present work, a methanol/air turbulent spray flame is studied using the transported pdf method, whereas presumed pdf modeling is performed for an ethanol/air spray flame. In this section, the results from both flames are presented and discussed. Experimental data is used to validate the numerical computations.

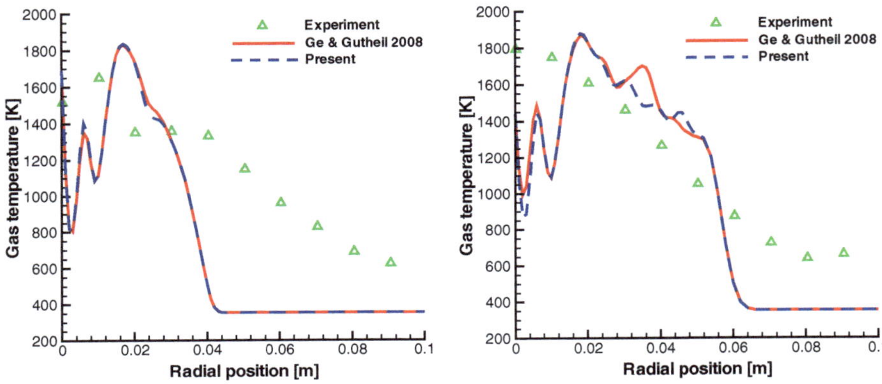

Fig. 1 Effect of the spray source modification on the gas temperature at cross sections of 0.1 m (*left*) and 0.15 m (*right*) from nozzle exit

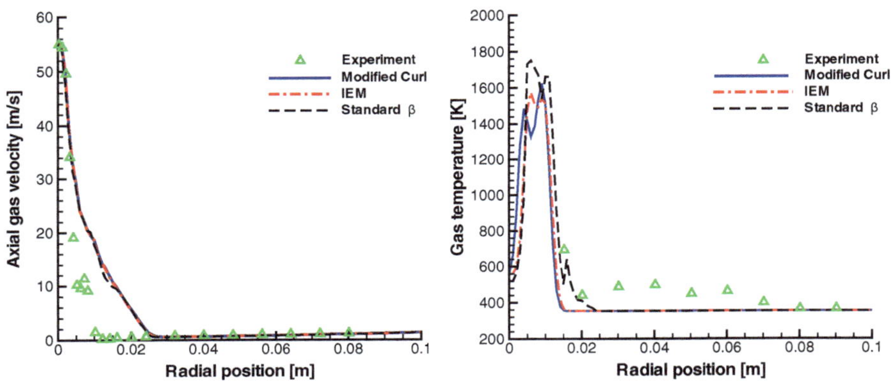

Fig. 2 Experimental and numerical profiles of the axial gas velocity (*left*) and the mean gas temperature (*right*) at the cross section of 0.025 m distance from the nozzle exit

3.1 Methanol/Air Spray Flame–Transported Pdf Modeling

A turbulent methanol/air spray flame is modeled using the transported joint mixture fraction–enthalpy pdf. The initial and boundary conditions for the simulations are taken from the experiment, which was conducted by McDonell and Samuelsen [31]. In the experiment, a dilute methanol spray is injected into a turbulent air flow with liquid inflow rate of 1.32 g/s, the air flow results in a pressure drop of 3.73 kPa. The gas velocities, droplet size distribution, gas temperature and concentration of methanol vapor are measured at the cross sections of 0.0075 m, 0.025 m, 0.05 m, 0.075 m, 0.1 m, and 0.15 m away from the nozzle exit. The gas velocity and droplet size are measured using phase Doppler interferometry (PDI). The concentration of the methanol vapor is obtained using infrared extinction/scattering (IRES). The experimental data at the cross section of 0.0075 m away from the nozzle exit are

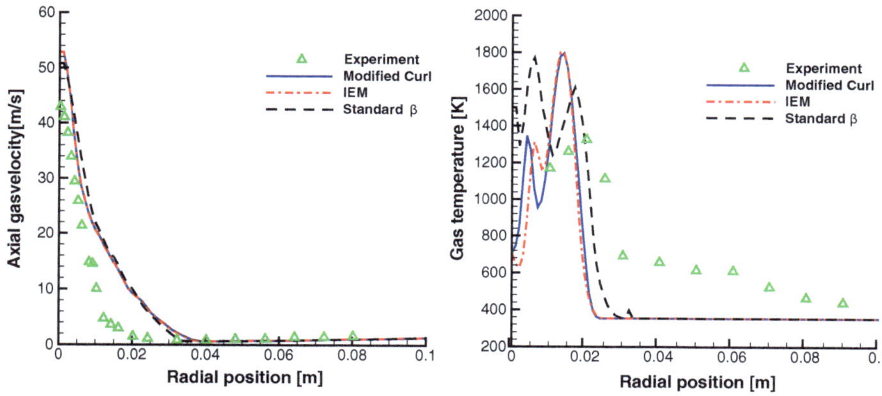

Fig. 3 Experimental and numerical profiles of the axial gas velocity (*left*) and the mean gas temperature (*right*) at the cross section of 0.05 m distance from the nozzle exit

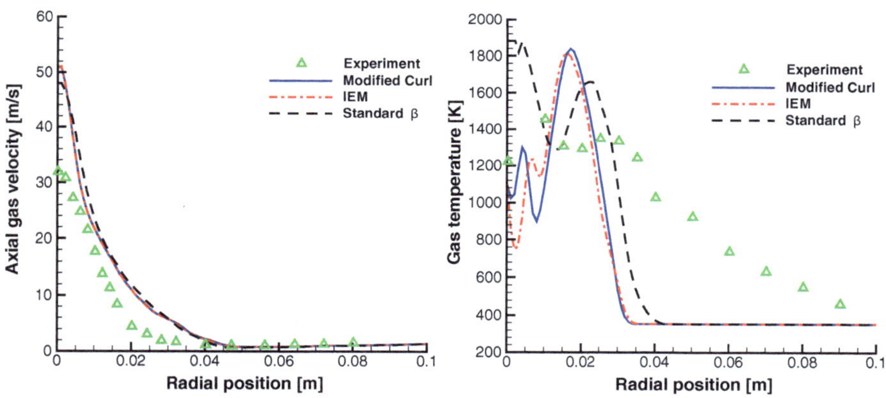

Fig. 4 Experimental and numerical profiles of the axial gas velocity (*left*) and the mean gas temperature (*right*) at the cross section of 0.075 m distance from the nozzle exit

used to generate the inlet profiles for the numerical computations, and the numerical results are compared with experimental data at the cross sections further downstream.

In the present work, the spray source terms for the formulation of the joint mixture fraction–enthalpy pdf are revised as derived in Sect. 2.2. The new formulation includes two new spray source terms appearing in Eq. 24 as well as the fluctuations of source terms as given in Eqs. 25 and 26. The fluctuations of the source terms are considered negligible, but the additional spray source terms are included. This leads to a revised formulation of the IEM model, where two additional spray source terms are obtained, see Eqs. 27 and 28. These new terms appear in closed form. A comparison of the previous computations and the new formulation is provided in Fig. 1, where their effect on the gas temperature profiles at cross sections of 0.1 m (left) and 0.15 m (right) away from nozzle exit is displayed. The differences between the new formulation (dashed lines) and the previous results (solid lines) are small

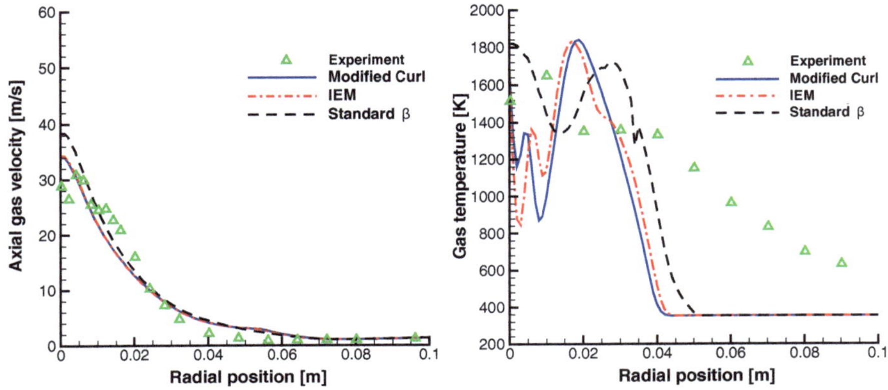

Fig. 5 Experimental and numerical profiles of the axial gas velocity (*left*) and the mean gas temperature (*right*) at the cross section of 0.1 m distance from the nozzle exit

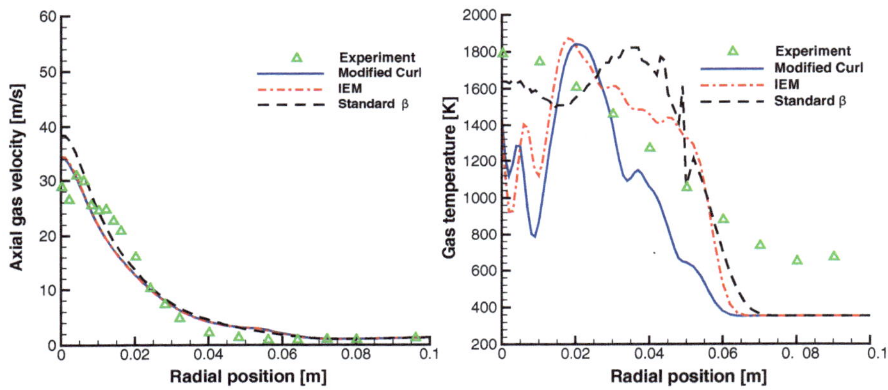

Fig. 6 Experimental and numerical profiles of the axial gas velocity (*left*) and the mean gas temperature (*right*) at the cross section of 0.15 m distance from the nozzle exit

for the present conditions with relatively low liquid mass load, but they may become more relevant if more dense sprays are considered.

The IEM model has been used extensively in the turbulent mixing model for gas combustion, and there, the modified Curl model has been discussed as suitable alternative to the IEM model. The modified Curl model is revised for use in turbulent spray combustion, see Eqs. 29–32. Figures 2–6 show the computed and the experimental profiles of the axial gas velocity and the mean gas temperature at various cross sections, where experimental data are available. Blue solid lines show results obtained with the modified Curl model and dash-dot red lines display computations with the IEM model. Simulations using the standard β distribution results are displayed using a black dashed line. Symbols present experimental data.

At the cross section of 0.025 m away from the nozzle exit, see Fig. 2, the numerical results of the gas velocity coincide near the centerline and in the outer jet whereas there is deviation at the jet boundary. All models perform similarly, and only marginal

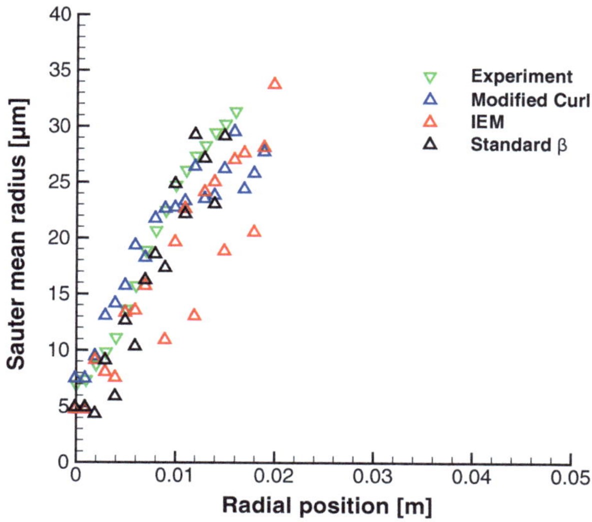

Fig. 7 Experimental and numerical profiles of the Sauter mean radius at the cross section of 0.025 m distance from the nozzle exit

differences are noticeable. At the highest cross section of 0.15 m away from the nozzle exit, see Fig. 6, the standard β function attains higher values than both other model results and the experiment at the centerline, so that the transported pdf models perform better than the presumed pdf approach as far as the gas velocity is concerned. The right part of Figs 2–6 show the profiles of the gas temperature, and there is larger deviation between numerical and experimental results. Experimental values of gas temperature are not available near the centerline, which makes the comparison of the models somewhat difficult. Moreover, the generation of initial profiles of gas temperature for the numerical simulations therefore has some uncertainties. At higher distances from the nozzle exit, the standard β function predicts too high values at the centerline where the transported pdf model shows too low values. It seems that the modified Curl model performs somewhat better than the IEM model in this region, see Fig. 4. Somewhat farther away from the centerline, however, all models underestimate the experimental results, so that a final conclusion about the performance of the mixing models cannot be drawn. In future, these models will be applied to simulate different turbulent spray flames [30], which may lead to a better conclusion. Note that the profiles of gas velocity are much less affected by the different models compared to the gas temperature, since the differences in IEM and modified Curl models act mainly on the species and gas temperature profiles.

The flamelet library consists of two different initial droplet radii of 10 μm and 25 μm. Therefore, the droplet characteristics are very important in describing the spray flames. Figure 7 shows the computed and experimental values of the Sauter mean radius at the cross section of 0.025 m from nozzle exit. It can be seen that the modified Curl model improves the results of the IEM and the presumed pdf method. In fact, the deviation between IEM and experiment is largest and most scattered. Therefore, it is concluded that the transported pdf approach with the modified Curl mixing model performs best with respect to the prediction of the Sauter mean droplet radius.

3.2 Ethanol/Air Spray Flame–Presumed Pdf Modeling

The spray burner for this ethanol flame was set up at University of Heidelberg and studied by Düwel et al. [12]. The burner consists of a pressure-atomizer nozzle (Delavan 67700-5) with a diameter of 10 mm, which is fixed about 80 mm above the center of a multi-hole brass plate. This fuel nozzle produces a hollow cone spray. The co-flowing air passes the glass balls, honeycomb grid, and the perforated plate to generate a homogeneous flow. A stable flame is obtained by pre-heating the ethanol to 318 K at the nozzle exit. The fuel pressure is varied between 1.4 and 2.6 bar. The resulting liquid flow rate varies between 0.39 and 0.54 g/s. The air coflow velocity is varied between 0 and 0.64 m/s. The liquid-phase temperature field is measured by two-color LIF thermometry. Droplet sizes are measured by LIF/Mie drop-sizing. Moreover, droplet size distributions and droplet velocities are measured with phase-Doppler anemometry (PDA). For the simulation, the measured droplet size and velocity distributions at 2 mm away from the nozzle exit are used as inlet conditions. The gas temperature at the inlet is approximated from the multi-line NO-LIF measurements. The gas velocity at the inlet is estimated from the total flux of the air coflow.

Since the transported pdf approach presented in Sect. 3.1 is very expensive because of the long computational time involved in the Monte Carlo simulations, the remainder of this chapter concerns the study of the four parameter presumed β function and, in particular, the choice of the two additional parameters $\xi_{\min}$ and $\xi_{\max}$. As described in Sect. 2.3, these parameters are considered to be associated with the variance or standard deviation of the mixture fraction, and a more detailed investigation of these variables is of interest. The contour plots of the mean, $\widetilde{\xi}$, and standard deviation, $\sqrt{\widetilde{\xi''^2}}$, of the mixture fraction are shown in Fig. 8, and its turbulence intensity, $\sqrt{\widetilde{\xi''^2}}/\widetilde{\xi}$, is presented in Fig. 9. The absolute values of the mean mixture fraction range up to about 0.1 for the stoichiometric ethanol/air spray flame. The main evaporation zone lies in the area of high mean values of the mixture fraction, whereas elevated values of the standard deviation reside near the centerline somewhat downstream of the fuel injector. The turbulence intensity shown in Fig. 9 attains relatively uniform values, and in the area of interest, they range between about 0.5 and 2. The highest values occur in regions of low or vanishing vales of the mean mixture fraction between a radial distance of 0.02 and 0.04 m near the nozzle exit, which are not of interest.

Figures 10–14 show the comparison of standard β presumed pdf with modified β presumed pdfs at different cross sections with different choices of $\xi_{\min}$ and $\xi_{\max}$. Note that the standard two-parameter β function is retrieved for the values of zero and unity for $\xi_{\min}$ and $\xi_{\max}$, the results of which are shown using solid lines. The left part of the figures displays the axial gas velocity, where no experimental data are available for comparison, whereas the right part of the figures show a comparison of the gas temperatures. Here, symbols represent experimental values. Both the minimum and maximum values defined in Eq. 43 are varied following Eqs. 51 and 52. The minimum

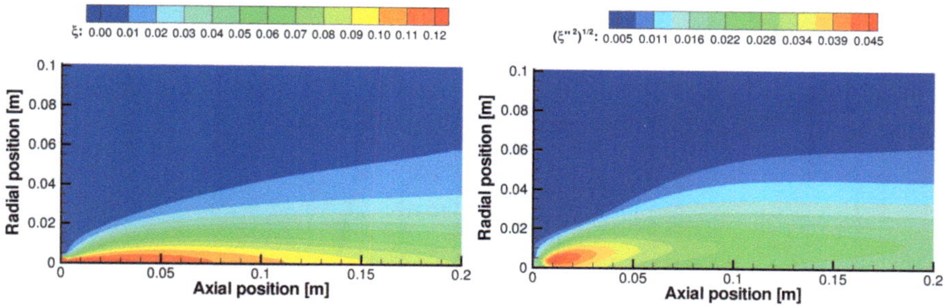

Fig. 8 Contour plot of the mean value (*left*) and the standard deviation (*right*) of the mixture fraction of the ethanol/air spray flame

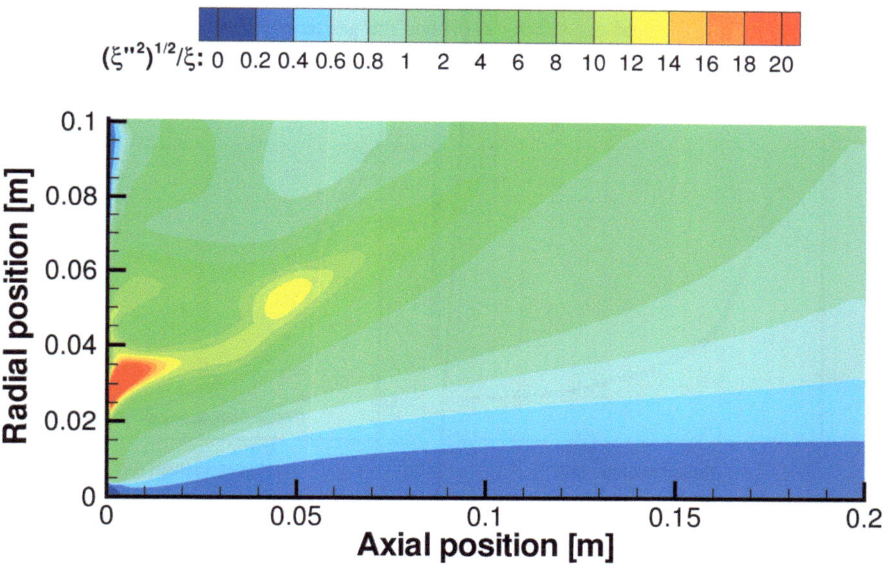

Fig. 9 Contour plot of the turbulence intensity of the ethanol/air spray flame

value is restrained by the lower limit of zero and the maximum value by the upper limit of unity. Since the stoichiometric value of the mixture fraction for ethanol/air combustion is near 0.1 and therefore, it lies close to the lower limit, only small values of c, see Eq. 51, may be evaluated. The present choice is $c \in \{1, 2\}$. The upper limit of unity for the maximum value lies well above the stoichiometric value of the mixture fraction, so that the parameter d in Eq. 52 is chosen as $d \in \{1, 2, 8\}$. In all situations, the restrictions

$$\xi_{\min} = \max\{0, \widetilde{\xi} - c\sqrt{\widetilde{\xi''^2}}\} \tag{53}$$

and

$$\xi_{\max} = \min\{\widetilde{\xi} + d\sqrt{\widetilde{\xi''^2}}, 1\} \tag{54}$$

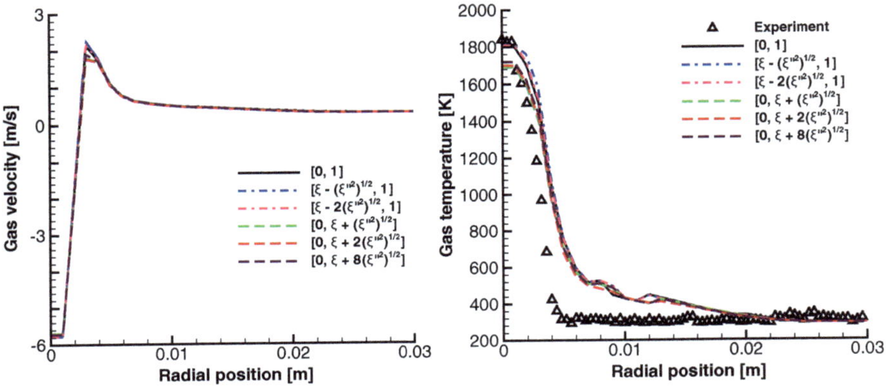

Fig. 10 Profiles of the axial gas velocity (*left*) and the gas temperature (*right*) at cross section 0.006 m from the nozzle exit

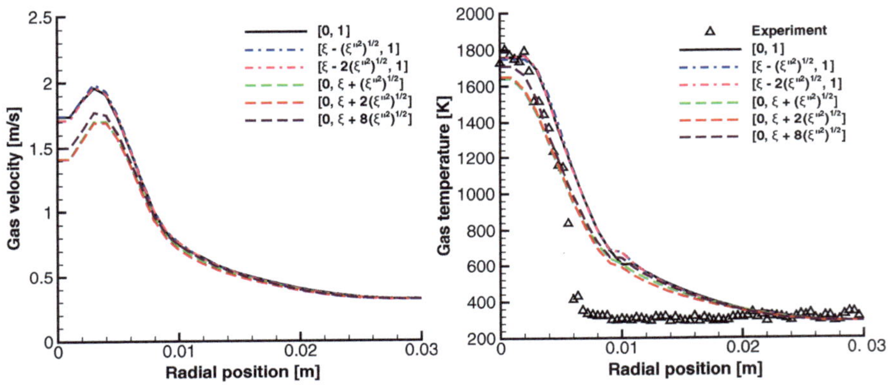

Fig. 11 Profiles of the axial gas velocity (*left*) and the gas temperature (*right*) at cross section 0.01 m from the nozzle exit

apply. The results shown in Figs. 10–14 refer to the turbulent ethanol/air spray flame at 0.006, 0.01, 0.02, 0.025 and 0.03 m away from the nozzle exit.

At the first cross section, see Fig. 10, the differences between the different models are small because the location is near the nozzle exit where the turbulent fluctuations are relatively small, c.f. right part of Fig. 8. A comparison of the modification of the value ξ_{min} with $c = 1$ or 2 shows that the differences here are minor compared to the standard β pdf. This is because of the low value of stoichiometric mixture fraction, which leads to the restriction given in Eq. 53 to become effective. This is confirmed by the fact that the curve with $c = 1$ lies further away from the two parameter β formulation than the four parameter β function for $c = 2$. A higher value for c will not lead to different results than those obtained by the standard β function. Near the centerline, the modification of ξ_{min} leads to somewhat smaller values compared to the standard β function, and at higher distances from the axis, higher values

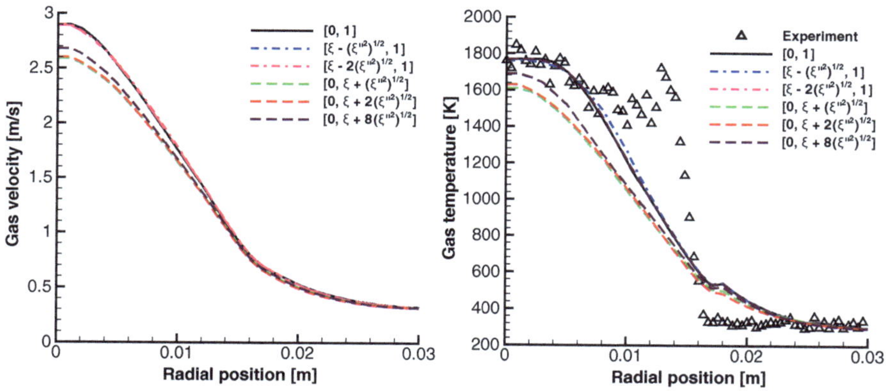

Fig. 12 Profiles of the axial gas velocity (*left*) and the gas temperature (*right*) at cross section 0.02 m from the nozzle exit

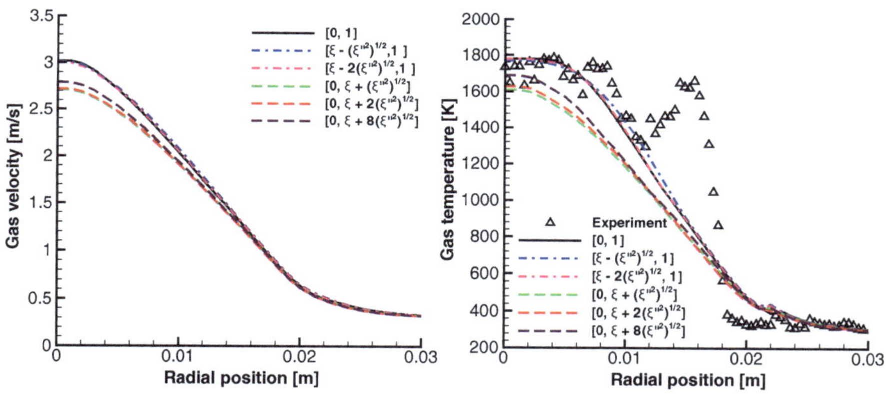

Fig. 13 Profiles of the axial gas velocity (*left*) and the gas temperature (*right*) at cross section 0.025 m from the nozzle exit

are computed. For the present situation of a stoichiometric flame with very low stoichiometric value of mixture fraction, it is concluded that the modification of ξ_{min} hardly affects the results, and the standard value zero may be retained. For fuel rich conditions, however, the choice may have a stronger effect on the results.

The variation of the maximum value has a stronger influence on the simulations, where the simulation with $d = 1$ shows the biggest difference compared to the standard β function. An increase of the parameter d causes the profile to approach that of the standard β function. At very high values of d, the condition given in Eq. 54 takes effect, and as the limit, the two-parameter function is approached. However, it is seen that the modification of the maximum value ξ_{max} leads to results, which lie below those of the standard β function.

For the profile of the gas temperature at 0.006 m distance from the nozzle exit, the choice of the value $d = 1$ improves the results near the centerline, whereas at

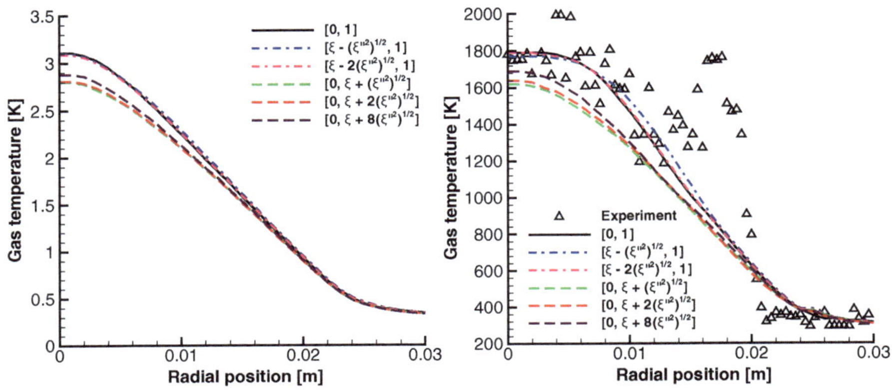

Fig. 14 Profiles of the axial gas velocity (*left*) and the gas temperature (*right*) at cross section 0.03 m from the nozzle exit

0.01 m, the value of 8 probably performs best. Beyond 0.01 m from the nozzle exit, the experimental results fluctuate, and it is difficult to judge which choice of the parameter d performs best, but the value of $d = 8$ is probably a good choice.

In summary, it can be concluded that the modification of the standard β function to a four parameter function, in general, may be a good choice, but the results of the modified β function are far away from the results obtained with the transported pdf [17] for the present choice of the new parameters, which needs reconsideration. ξ_{min} and ξ_{max} should possibly not (only) be linked to the standard deviation of the mixture fraction as assumed in the present study. However, for stoichiometric or lean combustion, a three parameter modified β function with $\xi_{min} = 0$ is a sufficient choice for the present experiment.

4 Conclusions

A methanol/air spray flame is modeled using a transported joint pdf equation for the mixture fraction and the enthalpy, which is derived in the present study. The unclosed term of molecular mixing is described using an extended IEM model, and an extended modified Curl model is proposed and applied. In particular, the modified Curl model is extended to account for the spray evaporation. Detailed chemistry is considered through a spray flamelet model [24]. The comparison between the transported joint pdf model with both mixing models and the standard β function and experiment is shown. Near the centerline, all models deviate from the experimental data, and at higher distances it is found that the transported pdf model with the extended modified Curl model performs best even though none of the models is able to capture the second peak in the profile of gas temperature. The different models also affect the profiles of the Sauter mean diameter, and the extended modified Curl model improves the results of the IEM model.

A presumed four parameter extended β function is used to model the turbulent ethanol/air spray flame. The two new parameters are assumed to be linked to the standard deviation of the mixture fraction. It appears that the minimum value introduced in the four parameter function hardly affects the numerical results of stoichiometric ethanol/air spray flame with its low values of Favre averaged mixture fraction. In this situation, a three parameter function is sufficient, i.e. $\xi_{\min} = 0$. A parameter study of the maximum value shows that for the present conditions, the value of $\xi_{\max} = \widetilde{\xi} + 8\sqrt{\widetilde{\xi''^2}}$ performs best. However, future studies should reconsider the assumed link of this parameter to the standard deviation of the mixture fraction.

Acknowledgements The authors gratefully acknowledge the financial support of Heidelberg School of Mathematical and Computational Sciences for their financial support. YH acknowledges funding through the China Scholarship Council.

References

1. Abramzon, B., Sirignano, W. A.: Droplet vaporization model for spray combustion calculation. Int. J. Heat Mass Transfer. **32**, 1605–1618 (1989).
2. Anand, G., Jenny, P.: Stochastic modeling of evaporating sprays within a consistent hybrid joint PDF framework. J. Comput. Phys. **228**, 2063–2081 (2009).
3. Atkins, P., Paula, J. D.: Atkins Physical Chemistry, Oxford Higher Education, 7th Edition, (2001).
4. Batchelor, G. K.: An Introduction to Fluid Dynamics. Cambridge University Press, London, (1967).
5. Cao, R. R., Wang, H., Pope, S. B.: The effect of mixing models in PDF calculations of piloted jet flames. Proc. Combust. Inst. **31**, 1543–1550 (2007).
6. Crowe, C. T., Sharma, M. P., Stock, D. E.: The particle-source-in cell (PSI-Cell) model for gas-droplet flows. J. Fluids Eng. **99**, 325–332 (1977).
7. Curl, R. L.: Dispersed phase mixing: 1. Theory and effects in simple reactor. AIChE J. **9**(2), 175–181 (1963).
8. De, S., Lakshmisha, K. N., Bilger, R. W.: Modeling of nonreacting and reacting turbulent spray jets using a fully stochastic separated flow approach. Combust. Flame. **158**, 1992–2008 (2011).
9. Demoulin, F. X., Borghi, R.: Assumed PDF modeling of turbulent spray combustion. Combust. Sci. Technol. **158**, 249–271 (2000).
10. Dopazo, O., O'Brien, E. E.: An approach to the auto-ignition of a turbulent mixture. Acta Astronaut. **1**, 1239–1266 (1974).
11. Durand, P., Gorokhovski, M., Borghi, R.: An application of the probability density function model to diesel engine combustion. Combust. Sci. Technol. **144**, 47–78 (1999).
12. Düwel, I., Ge, H.-W., Kronemayer, H., Dibbel, R., Gutheil, E., Schulz, C.: Experimental and numerical characterization of a turbulent spray flame. Proc. Combust. Inst. **31**, 2247–2255 (2007).
13. Eckstein, J., Chen, J. Y., Chou, C. P., Janicka, J.: Modeling of turbulent mixing in opposed jet configuration: one-dimensional Monte Carlo probability density function simulation. Proc. Combust. Inst. **28**, 141–148 (2000).
14. Garg, R., Narayanan, C., Subramaniam, S.: A numerically convergent Lagrangian-Eulerian simulation method for dispersed two-phase flows. Int. J. Multiphase Flow. **35**, 376–388 (2009).
15. Ge, H.-W., Gutheil, E.: PDF simulation of turbulent spray flows. Atomization Sprays. **16**, 531–542 (2006).

16. Ge, H.-W., Gutheil, E.: Simulation of a turbulent spray flame using coupled PDF gas phase and spray flamelet modeling. Combust. Flame. **153**, 173–185 (2008).
17. Ge, H.-W., Düwel, I., Kronemayer, H., Dibble, R. W., Gutheil, E., Schulz, C., Wolfrum, J.: Laser based experimental and Monte Carlo PDF numerical investigation of an ethanol/air spray flame. Combust. Sci. Technol. **180**, 1529–1547 (2008).
18. Ge, H.-W., Hu, Y., Gutheil, E.: Joint gas-phase velocity-scalar PDF modeling for turbulent evaporating spray flows. Combust. Sci. Technol. **184**, 1664–1679 (2012).
19. Gordon, R. L., Masri, A. R., Pope, S. B., Goldin, G. M.: A numerical study of auto-ignition in turbulent lifted flames issuing into vitiated co-flow. Combust. Theor. Model. **11**, 351–376 (2007).
20. Gutheil, E.: Structure and extinction of laminar ethanol-air spray flames. Combust. Theor. Model. **5**(2), 131–145 (2001).
21. Gutheil, E., Sirignano, W. A.: Counterflow spray combustion modeling with detailed transport and detailed chemistry. Combust. Flame. **113**, 92–105 (1998).
22. Heye C. R., Raman, V., Masri, A. R.: LES/probability density function approach for the simulation of an ethanol spray flame. Proc. Combust. Inst. **34**, 1633–1641 (2013).
23. Hollmann, C., Gutheil, E.: Modeling of turbulent spray diffusion flames including detailed chemistry. Proc. Combust. Inst. **26**, 1731–1738 (1996).
24. Hollmann, C., Gutheil, E.: Flamelet-modeling of turbulent spray diffusion flames based on a laminar spray flame library. Combust. Sci. Technol. **135**, 175–192 (1998).
25. Hubbard, G. L., Denny, V. E., Mills, A. F.: Droplet evaporation: effects of transient and variable properties. Int. J. Heat Mass Transfer. **18**, 1003–1008 (1975).
26. Kung, E. H., Haworth, D. C.: Transported probability density function (tPDF) modeling for direct-injection internal combustion engines. SAE Paper 2008-01-0969; SAE Int. J. Engines. **1**, 591-606 (2009).
27. Liu, Z. H., Zheng, C. G., Zhou, L. X.: A joint PDF model for turbulent spray evaporation/combustion. Proc. Combust. Inst. **29**, 561–568 (2002).
28. Lundgren, T. S.: Model equation for non-homogeneous turbulence. Phys. Fluids. **12**, 485–497 (1969).
29. Luo, K., Pitsch, H., Pai, M. G., Desjardins, O.: Direct numerical simulations and analysis of three dimensional n-heptane spray flames in a model swirl combustor. Proc. Combust. Inst. **33**, 2143–2152 (2011).
30. Masri, A., Gounder, J.: Details and Complexities of Boundary Conditions in Turbulent Piloted Dilute Spray Jets and Flames. In: Bart, Merci, Dirk, Roekaerts and Amsini, Sadiki (Eds.), Experiments and Numerical Simulations of Diluted Spray Turbulent Combustion, 41–68, New York, Springer, (2011).
31. McDonell, V. G., Samuelsen, G. S. : An experimental data base for computational fluid dynamics of reacting and nonreacting methanol sprays. J. Fluids Eng. **117**, 145–153 (1995).
32. Mehta R. S.: Detailed Modeling of soot formation and turbulence-radiation interactions in turbulent jet flames. Ph.D. thesis, The Pennsylvania State University, (2008).
33. Menon, S., Fureby, C.: Computational Combustion. In: Encyclopedia of Aerospace Engineering, Wiley, (2010).
34. Miller, R. S., Bellan, J.: On the validity of the assumed probability density function method for modeling binary mixing/reaction of evaporated vapor in gas-liquid turbulent shear flow. Proc. Combust. Inst. **27**, 1065–1072 (1998).
35. Mortensen M., Bilger R.: Derivation of the conditional moment closure equation for spray combustion. Combust. Flame. **156**, 62–72 (2009).
36. Naud, B.: PDF modeling of turbulent sprays and flames using a particle stochastic approach. Ph. D. Thesis, TU Delft, (2003).
37. Olguin, H., Gutheil, E.: Influence of evaporation on spray flamelet structures, Combust. Flame (2013), http://dx.doi.org/10.1016/j.combustflame.2013.10.010.
38. Pai, G. M., Subramaniam, S.: A comprehensive probability density function formalism for multiphase flows. J. Fluid Mech. **628**, 181–228 (2009).

39. Peters, N.: Laminar diffusion flamelet models in non-premixed turbulent combustion. Prog. Energy Combust. Sci. **10**, 319–339 (1984).
40. Pope, S. B.: The relationship between the probability approach and particle models for reaction in homogeneous turbulence. Combust. Flame. **35**, 41–45 (1979).
41. Pope, S. B.: Transport equation for the joint probability density function of velocity and scalars in turbulent flow. Phys. Fluids. **24**, 588–596 (1981).
42. Pope, S. B.: PDF methods for turbulent reactive flows. Prog. Energy Combust. Sci. **11**, 119–192 (1985).
43. Raju, M. S.: Application of scalar Monte Carlo probability density function method turbulent spray flames. Numer. Heat Transfer A. **30**, 753–777 (1996).
44. Richardson, J. M., Howard, H. C., Smith, R. W.: The relation between sampling-tube measurements and concentration fluctuations in a turbulent gas jet. Proc. Combust. Inst. **4** 814–817, (1953).
45. Rumberg, O., Rogg, B.: Full PDF modeling of reactive sprays via an evaporation-progress variable. Combust. Sci. Technol. **158**, 211–247 (2000).
46. Schiller, L., Neumann, A. Z.: A drag coefficient correlation. VDI Zeitschrift **77**, 318–320 (1933).
47. Wang, H., Pope, S. B.: Lagrangian investigation of local extinction, re-ignition and auto-ignition in turbulent flames. Combust. Theory Modeling. **12**, 857–882 (2008).
48. Zhu, M., Bray, K. N. C., Rumberg, O., Rogg, B.: PDF transport equations for two-phase reactive flows and sprays. Combust. Flame. **122**, 327–338 (2000).

Index

B. Merci, E. Gutheil (eds.), *Experiments and Numerical Simulations of Turbulent Combustion of Diluted Sprays*, ERCOFTAC Series 19, DOI 10.1007/978-3-319-04678-5,
© Springer International Publishing Switzerland 2014

MIX
Papier aus verantwortungsvollen Quellen
Paper from responsible sources
FSC® C105338

If you have any concerns about our products,
you can contact us on
ProductSafety@springernature.com

In case Publisher is established outside the EU,
the EU authorized representative is:
Springer Nature Customer Service Center GmbH
Europaplatz 3, 69115 Heidelberg, Germany

Printed by Libri Plureos GmbH
in Hamburg, Germany